Introdu[illegible]ion [illegible] Semiconductor Physics

Semiconductor Electronics Education Committee Books

Vol. 1 ***Introduction to Semiconductor Physics***
R. B. Adler, A. C. Smith, and R. L. Longini

Vol. 2 ***Physical Electronics and Circuit Models of Transistors***
P. E. Gray, D. DeWitt, A. R. Boothroyd, and J. F. Gibbons

Vol. 3 ***Elementary Circuit Properties of Transistors***
C. L. Searle, A. R. Boothroyd, E. J. Angelo, Jr., P. E. Gray, and D. O. Pederson

Vol. 4 ***Characteristics and Limitations of Transistors***
R. D. Thornton, D. DeWitt, E. R. Chenette, and P. E. Gray

Vol. 5 ***Multistage Transistor Circuits***
R. D. Thornton, D. O. Pederson, C. L. Searle, E. J. Angelo, Jr., and J. Willis

Vol. 6 ***Digital Transistor Circuits***
J. N. Harris

Vol. 7 ***Handbook of Basic Transistor Circuits and Measurements***
R. D. Thornton, J. G. Linvill, E. R. Chenette, H. L. Ablin, J. N. Harris, and A. R. Boothroyd

Introduction to Semiconductor Physics

Semiconductor Electronics Education Committee, Volume 1

R. B. Adler
Massachusetts Institute of Technology

A. C. Smith
Massachusetts Institute of Technology

R. L. Longini
Carnegie Institute of Technology
formerly with Westinghouse Research Laboratories,
Pittsburgh, Pa.

John Wiley & Sons, Inc., New York · London · Sydney

9 10

Library of Congress Catalog Card Number: 64-23824

Printed in the United States of America

ISBN 0 471 00887 7

Foreword

The importance of transistors and other semiconductor devices is now well established. The subsequent development of microminiaturized electronic circuits has blurred the dividing line between the "device" and the "circuit," and thus has made it increasingly important for us to understand deeply the relationship between the internal physics and structure of a device, and its potentialities for circuit performance. Furthermore, the small size and efficient operation of semiconductor devices make possible for the first time a much closer integration between the theoretical and laboratory aspects of the educational process.

To prepare new educational material which would reflect these developments, there was formed in the Fall of 1960 a group known as the Semiconductor Electronics Education Committee (SEEC). This committee is comprised of university and industrial members, brought together by several of the faculty of the Electrical Engineering Department at the Massachusetts Institute of Technology, with Professor C. L. Searle acting as Chairman and Professor R. B. Adler acting as Technical Director. The committee undertook the production of a multipurpose course in semiconductor electronics, designed primarily for use in universities at the third or fourth year undergraduate level.

The success of the high-school physics course developed by the Physical Science Study Committee (PSSC) lead the SEEC to believe that the same kind of combination used there—text, laboratory experiments, and films, in a complementary format—would be the most practical way of providing uniformly high-quality instruction over the wide range of material involved. It was hoped that this arrangement would lead to broad applicability of the course in the academic world, and also in some professional training activities of industry and government. This book is one in the SEEC series, all volumes of which are listed here:*

Vol. 1 (ISP) *Introduction to Semiconductor Physics,* R. B. Adler, A. C. Smith, and R. L. Longini
Vol. 2 (PEM) *Physical Electronics and Circuit Models of Transistors,* P. E. Gray, D. DeWitt, A. R. Boothroyd, and J. F. Gibbons
Vol. 3 (ECP) *Elementary Circuit Properties of Transistors,* C. L. Searle, A. R. Boothroyd, E. J. Angelo, Jr., P. E. Gray, and D. O. Pederson
Vol. 4 (CLT) *Characteristics and Limitations of Transistors,* R. D. Thornton, D. DeWitt, E. R. Chenette, and P. E. Gray
Vol. 5 (MTC) *Multistage Transistor Circuits,* R. D. Thornton, D. O. Pederson, C. L. Searle, E. J. Angelo, Jr., and J. Willis
Vol. 6 (DTC) *Digital Transistor Circuits,* J. N. Harris
Vol. 7 (TCM) *Handbook of Basic Transistor Circuits and Measurements,* R. D. Thornton, J. G. Linvill, E. R. Chenette, H. L. Ablin, J. N. Harris, and A. R. Boothroyd

These books have all gone through at least one "preliminary edition," many through two or more. The preliminary editions were used in teaching trials at some of the participating colleges and industrial training activities, and the results have been used as a basis for revision.

*Minor changes in title or authorship may take place in some of the volumes, which are still in preparation at the time of this writing.

It is almost impossible to enumerate all those people who have contributed some of their effort to the SEEC. Certain ones, however, have either been active with the Committee steadily since its inception, or have made very major contributions since then. These may be thought of as "charter members," deserving special mention.

From Universities

California, Berkeley: D. O. Pederson
Imperial College, London: A. R. Boothroyd[Δ]
Iowa State: H. L. Ablin*
M.I.T.: R. B. Adler, P. E. Gray, A. L. McWhorter, C. L. Searle, A. C. Smith, R. D. Thornton, J. R. Zacharias, H. J. Zimmermann (Research Laboratory of Electronics), J. N. Harris (Lincoln Laboratory)
Minnesota: E. R. Chenette
New Mexico: W. W. Grannemann
Polytechnic Institute of Brooklyn: E. J. Angelo, Jr.
Stanford: J. F. Gibbons, J. G. Linvill
U.C.L.A.: J. Willis

From Industries

Bell Telephone Laboratories: J. M. Early, A. N. Holden, V. R. Saari
Fairchild Semiconductor: V. R. Grinich
IBM: D. DeWitt
RCA: J. Hilibrand, E. O. Johnson, J. I. Pankove
Transitron: B. Dale,† H. G. Rudenberg‡
Westinghouse Research Laboratories: A. I. Bennett, H. C. Lin, R. L. Longini§

General management of the SEEC operations is in the hands of Educational Services, Inc. (abbreviated ESI), Watertown, Mass.,

Δ Now at Queen's University, Belfast.
* Now at the University of Nebraska, Department of Electrical Engineering.
† Now at Sylvania Corp.
‡ Now at A. D. Little, Inc.
§ Now at Carnegie Institute of Technology, Department of Electrical Engineering.

a nonprofit corporation that grew out of the PSSC activities and is presently engaged in a number of educational projects at various levels. In addition to providing general management, ESI has supplied all the facilities necessary for preparing the SEEC films. These are 16-mm sound films, 20 to 40 minutes in length, designed to supplement the subject matter and laboratory experiments presented in the various text books. Two of the films are already available:

"Gap Energy and Recombination Light in Germanium"—J. I. Pankove and R. B. Adler

"Minority Carriers in Semiconductors"—J. R. Haynes and W. Shockley

At this writing two more films are still in the early stages of preparation: one deals with the *pn* junction, and the other with the relationship between physical structure, fabrication processes, and circuit performance of transistors. Pending arrangements for commercial distribution, completed films are available (purchase or rental) directly from Educational Services, Inc., 47 Galen Street, Watertown, Mass.

The committee has also endeavored to develop laboratory materials for use with the books and films. This material is referred to in the books and further information about it can be obtained from ESI.

The preparation of the entire SEEC program, including all the books, was supported at first under a general grant made to the Massachusetts Institute of Technology by the Ford Foundation, for the purpose of aiding in the improvement of engineering education, and subsequently by specific grants made to ESI by the National Science Foundation. This support is gratefully acknowledged.

Campbell L. Searle
Chairman, SEEC
Richard B. Adler
Technical Director, SEEC

Preface

An *Introduction to Semiconductor Physics* would take very different forms, according to the purposes and the audience for which it was prepared. This volume has been conceived primarily for use by persons who have a strong interest in electronic circuits, although no such circuits appear in the text. The connection between this book and electronic circuits therefore lies entirely in the selection of topics and in the choice of the extent to which each is treated quantitatively.

The topics have been selected to supply the minimum physical background which we believe is necessary to make the reader feel really comfortable with those parameters of a semiconductor that bear directly upon its performance in semiconductor devices. Even more specifically, we have stressed those matters related to transistors, without even attempting to give "fair weight" to other related devices. Our principal justification for the limitation is the growing need we find for a work (or series of works) that will provide students with the complete experience of being familiar enough with the internal physics of a device, to begin to apply it with a little sophistication in the design sense. But in addition, it has so far proved to be true that a thorough understanding of transistors makes easy the comprehension of all the other semiconductor devices.

Whereas the present book will, we are sure, stand on its own feet, it is only fair to point out that it was conceived as part of a short series designed for the over-all objective defined above. Inasmuch as this objective does not usually characterize a first course in electronics, use of the entire series of volumes was envisioned primarily for a "second contact" with electronics in general. By providing several separate volumes, like this one, however, we hope to have achieved enough flexibility for very wide use of the material, in whole or in part, from the junior year in the university through to the early graduate and professional level.

The extent of the physical background required for a satisfactory comprehension of a device and its applications is a matter open to considerable difference of opinion. There is little doubt about which concepts are needed, but there is a severe question about how much should be left to qualitative understanding and how much must be handled quantitatively. We feel that the bare *presentation* of formulas relating to the physics of the material or the device, without *any* attempt at proof, would be inconsistent with our objectives. A formula leads to an answer, not necessarily to any understanding. Accordingly, formulas without proof make sense only in situations where answers are more important than comprehension. Such is certainly not the case with respect to the undergraduate needs of either the potential circuit designer or device designer for knowledge of semiconductor physics.

Thus we have certainly tried to provide very careful qualitative discussions of all the significant points; but, being also believers in the precept that science and technology must be quantitative before they are really understood, we have also provided a quantitative treatment wherever possible. In some cases, the difference in space and time required by the quantitative discussion, as compared to the qualitative one, has been so great that we have placed the more elaborate argument in a special section, printed in small type size, which can be omitted without destroying the continuity of the text.

We believe that the objective of the whole program of which this volume is a part requires a close tie between the text material to be studied and positive learning action to be taken by the student. For this reason, the Problems and Laboratory Experiments have, in effect, been merged with the text by making coded references to them directly at points where their consideration is especially appropriate. The problem references are in some instances designed especially to provide practice with the related concepts, and in other instances to extend either the concepts or the quantitative development beyond their treatment in the text. The laboratory references serve the double role of relating the text material to its experimental foundations, and of providing a suitable detailed indication of the body of theory required as background for each experiment.

The concepts that this volume must establish to set the stage for the internal electronics of transistors are simply the following:

(*a*) The possibility of two distinct modes of electrical conduction associated with electron motion in a solid, leading to the concepts of *holes* and *conduction electrons* as *current carriers*.

(*b*) The various metallurgical, electrical, and environmental means of varying the number of each kind of carrier present in a semiconductor.

(*c*) The dynamical properties of the carriers, leading to the processes of *drift* in an electric field, *diffusion* in a concentration gradient, and mutual *generation* and *recombination* of the two types.

It would perhaps be pleasant if these few ideas could be disposed of in a few corresponding short paragraphs. Experience seems to prove that such a short discussion—often given in a first electronics course—becomes inadequate very rapidly, not only in respect to specific problems encountered in the applications, but also with respect to the confidence of the individual in his own capacity to deal securely with the devices in any nontrivial design situation.

Besides a previous course in elementary electronic-circuit ideas, the reader is assumed to have been exposed to enough modern physics to make him reasonably familiar with the notions of deBroglie wave mechanics, the Bohr picture of an atom, and light in the form of photons. The atomic or modern physics that often comes near the end of a regular two-year college physics course should be sufficient.

The authors of this volume owe to the authors of all other volumes in the series, and to the entire SEEC membership, a debt that is had to describe accurately. From the outside, it must be hard to believe that such a large "committee" could really function at a detailed level. Knowing the people involved, however, makes it obvious that strong effective participation was in fact bound to result.

Special acknowledgment must, however, go to Brian Dale, who provided the basis of a great deal of the laboratory material included in this book, and to Allan I. Bennett, Jr., for originating and developing the energy-gap experiment, and for numerous other laboratory suggestions.

We are also greatly indebted to Lawrence Castro, Richard Chang, Charles Counselman, Huber Graham, Lansing Hatfield, Roger Sudbury, Alton Tripp, and Wai Hon Lee, who as M. I. T. students worked with us during the various SEEC workshops on the experimental, numerical, and other detailed aspects of the task.

It is harder to give fair acknowledgment to all those students and staff members in this country and abroad whose comments, both written and oral, helped us through the two successive revisions of preliminary editions of this book. We hope that the long hours of discussion of the comments, and of editing and correcting the ideas, have improved the quality of the work as much as it has taught the writers.

Richard B. Adler
Arthur C. Smith
Richard L. Longini

Cambridge, Mass.
August 1964

Contents

* The portions of this section set in small type size may be omitted without destroying the continuity of text.

Introduction to Semiconductor Physics

1

The Valence-Bond Model of a Semiconductor

1.0 INTRODUCTION

We are here going to concern ourselves with materials known as semiconductors, primarily because their electrical properties are fundamental to the operation of transistors and related devices.

There are two different ways of thinking about the central problems of electrical conduction in semiconductors. The way we shall consider first, in this chapter, relies on the notion of a valence bond, which is relatively familiar from chemistry and which presents a visual picture of the conduction mechanism. This picture has the merit of being vivid enough to be remembered easily, although it can lead to dangerous byways of thought if it is taken too literally.

Unfortunately, the valence-bond point of view tends to be as weak for quantitative purposes as it is strong for qualitative ones. Therefore, in the next two chapters, we shall fill in the quantitative details in terms of another description, known as the energy-band picture. It is true that a certain amount of duplication of effort is bound to be involved when we look in two different ways at the same phenomenon; but it can also be argued with great force that our understanding is always incomplete if it is based on only a single viewpoint. In the present case, the two descriptions are complementary rather than duplicative, inasmuch as the first one emphasizes the events in space and time, whereas the second emphasizes them in energy and momentum.

1.1 ATOMS AND THEIR ARRANGEMENT IN MATTER

Matter, in general, is composed of atoms which are made up of positively charged nuclei surrounded by just enough electrons to make the complete atoms electrically neutral, as illustrated in Fig. 1.0*a*. The electrons are arranged in orbits, encircling each nucleus as planets encircle the sun. Distinctions between the chemical properties of different materials arise because they are

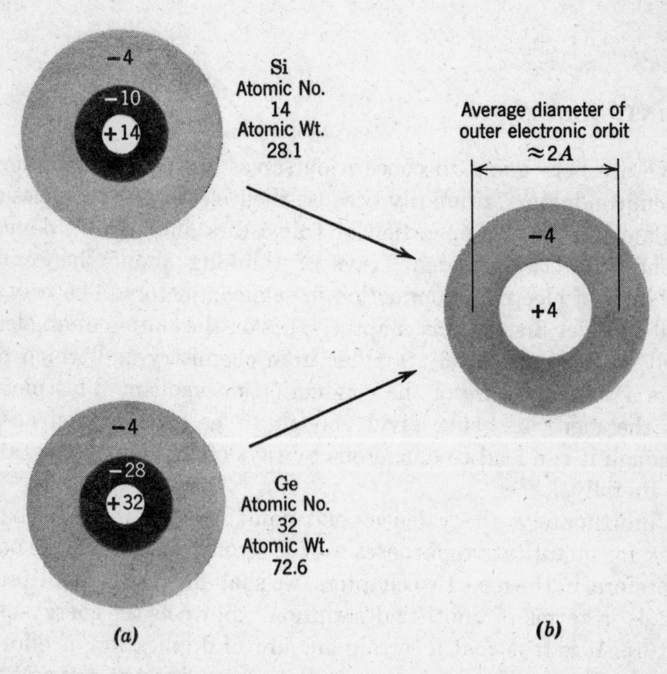

Fig. 1.0. Atoms (nuclei greatly oversize). (*a*) Schematic pictures of silicon and germanium, showing nuclei, electronic-orbit regions, and electrical neutrality. (*b*) Simplified schematic, in terms of an "effective nucleus" and valence electrons only.

composed of different atoms. Distinctions between various phases (solid, liquid, gaseous) of a given substance arise from differences in the strength with which its atoms are bound together. Increases in temperature produce increases in random thermal motion and corresponding increases in average atomic separation. Consequently, the average interatomic force is reduced.

Thus, if the atoms are far apart compared to the diameters of their outer electronic orbits (Fig. 1.0*b*), they cannot exert strong forces upon one another and therefore form a gas. If the separation between the atoms is decreased enough to become comparable to the diameters of their outer orbits, they may form a liquid or a solid. When the atomic separation gets small enough to make the orbits overlap considerably, interatomic forces become quite strong; the outermost electrons, familiar from chemistry as the valence electrons, may then play a very important part in determining both the interatomic forces and the final relative configuration of the various atoms.

When the atoms form a solid, they most often take up a three-dimensional orderly array that defines a crystal structure. Such a crystal structure is *periodic* in the sense that it may be translated parallel to itself in each of three independent directions by integer multiples of an appropriate distance on the atomic scale, and still look the same as it did before the translation. The minimum translation distances in each direction define a *unit cell* or building block, which is repeated throughout space to make the *crystal lattice* of the material. Since it is primarily with such crystalline solids that we shall be concerned, some of their special characteristics should be emphasized:

(1) Outer electronic orbits of isolated atoms generally have radii of about an angstrom (1 angstrom = 1A = 10^{-8} cm). Consequently, we should expect the atoms of a crystal to be separated by about twice this distance.

(2) Inasmuch as the outer electronic orbits of neighboring atoms would overlap considerably at the separations mentioned in (1), these orbits are significantly distorted by the relative proximity of the atoms in the crystal. In fact, it generally becomes difficult to say which electron goes with which atom, and some valence electrons are "shared" by several atoms.

(3) The regular crystal array of atoms implies that the valence electrons also adopt some systematic pattern of motion that joins each atom to its neighbors.

The formation of a crystal is like a three-dimensional jigsaw puzzle, in which each piece (atom) has a given number and symmetry of teeth (valence electrons), and the pieces must be fitted together to form a nice systematic array. One "rule of the game" is that the whole crystal must remain electrically neutral; another is that the best solutions to the puzzle are those structures which have minimum energy—for they are the stable configurations. Three simple examples of crystal arrangements are shown in Fig. 1.1 (P1.1, P1.2).†

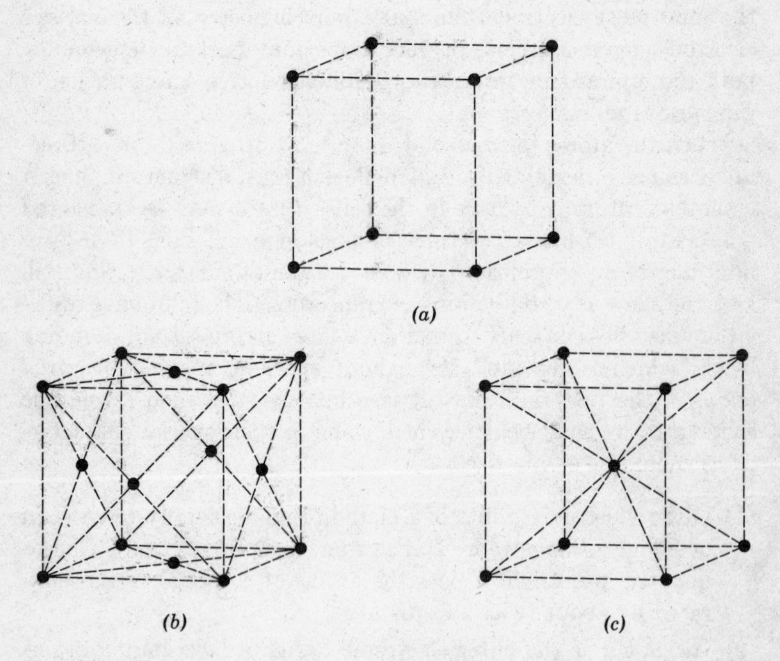

Fig. 1.1. Three examples of atoms arranged in a crystal. (*a*) Simplest cubic structure. (*b*) Face-centered cubic (FCC) structure. (*c*) Body-centered cubic (BCC) structure.

† At places in the text where it is suggested that particular Problems or Laboratory experiments be taken up, a suitable parenthetical reference is made. The Problems referred to appear at the end of each chapter, and the Laboratory experiments appear in the Appendix.

1.2 CONDUCTORS, INSULATORS, AND SEMICONDUCTORS

The crystal structure of common highly conductive metals, like copper, aluminum, and silver, is such that the outer atomic electrons are "shared" by *all* the atoms, as illustrated schematically in Fig. 1.2. These electrons are actually free to wander throughout the substance. This remains true over a very wide temperature range. In most metals, each atom supplies one such free electron, so that the number of free electrons per cubic centimeter is of the order of 10^{23}. Electrical conduction then takes place as a result of the motion of the free electrons under the action of an applied electric field.

In contrast with good conductors, the structure of solid insulators is such that over a wide temperature range almost all the electrons remain bound to the constituent atoms. Consequently, not many charges are available to move through an insulator with *moderate* electric fields applied to it, which means that it can have no appreciable electrical conductivity. The resistivity of good electrical insulators, like quartz, may be as high as 10^{18} ohm-cm, compared to 10^{-5} or 10^{-6} ohm-cm for a typical metal.†

From one point of view, semiconductors at room temperature are both poor conductors and poor insulators. Their resistivities at 300°K may lie in the range from 10^{-3} to 10^{5} ohm-cm. More precisely, we shall see that they are insulators at very low temperatures and become rather good conductors at high temperatures. The elements germanium and silicon are now among the most

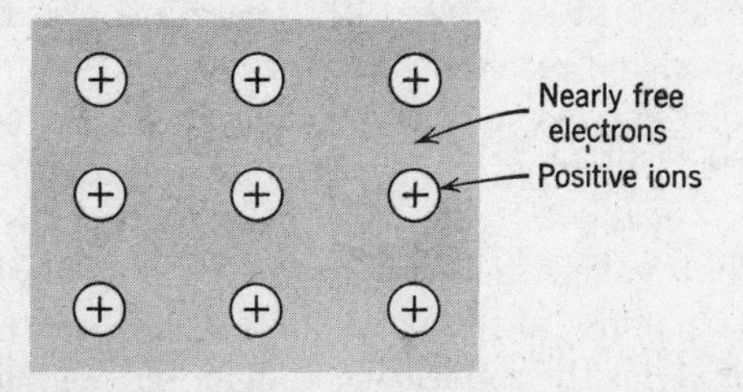

Fig. 1.2. Metal crystal, showing "sea" of shared electrons which neutralizes the positive crystal ions.

† The basic self-consistent system of units employed in this book is the *mks rationalized* one. But it is common in the semiconductor literature to depart from this system by using cm in place of m, especially for quantities connected with resistance or conductance.

important semiconductors used in electronics. They are brittle metallic-appearing crystalline solids that are used in almost all semiconductor diodes, although other materials are finding increasing use in these applications. For the moment, however, germanium and silicon appear to be the only materials really suitable for transistors. As examples of the semiconductor principles being discussed here, we shall use primarily these two elements. Some specific properties of these elements, along with some important general physical constants, appear in Table 1.0.

1.3 INTRINSIC CONDUCTION, CONDUCTION ELECTRONS, AND HOLES

1.3.1 *Valence Bonds*

In extremely "pure" form, germanium and silicon exhibit over a very wide temperature range what is known as intrinsic conductivity. By intrinsic conductivity, we mean electrical conduction by means of electrons which are present on account of the parent crystalline material only (Ge or Si in the present discussion), and not because of any foreign elements. When we speak thus of "purity," we mean in practice that not more than about 1 part in 10^{10} of the material is composed of foreign elements.

To understand the nature of the electrical conductivity of such pure material, it is important to realize that the crystal structure of germanium and silicon is the same as that of diamond (crystalline carbon). These elements are in the cubic crystal class of group IV of the periodic table (Fig. 1.3). These are also referred to as valence crystals, for reasons which will soon become clear. The essential features of the so-called *diamond structure* crystal appear in Fig. 1.4 (P1.3). Each atom is surrounded by four nearest neighbors, as shown in the dotted cube. Every atom tends to share one of its own four valence electrons with each of its four nearest-neighbor atoms, and also to take a share of one electron from each such neighbor. Thus the connecting bars in Fig. 1.4 may be thought of as tracks along which two electrons can shuttle back and forth in opposite directions between the associated atoms. Such an arrangement of shared electron pairs is called a *covalent bond*, or simply a *valence bond*. These bonds are the important cement that

TABLE 1.0

Some specific properties of Ge and Si, and some fundamental physical constants

Specific Properties†	Ge	Si
Atomic number	32	14
Atomic weight	72.60	28.06
Density (25°C), kg/m^3	5.33×10^3	2.33×10^3
Melting point, °C	936	1420
Boiling point, °C	2700	2600
Thermal conductivity (25°C), watts/m°C	63	84
Specific heat (25°C), joules/kg°C	310	760
Linear thermal expansion coefficient, °C^{-1}	6.1×10^{-6}	4.2×10^{-6}
Relative dielectric constant, ϵ/ϵ_o	16	12
Effective mass ratio $\frac{m^*}{m}$ (from density of states at 4°K):		
conduction electrons	0.55	1.1
holes	0.37	0.59

Fundamental Physical Constants‡	
Electron charge q, coulomb	1.60×10^{-19}
Electron rest mass m, kg	9.11×10^{-31}
Proton rest mass, kg	1.67×10^{-27}
Planck's constant h, joule-sec	6.62×10^{-34}
Boltzmann's constant k, joule/°K	1.38×10^{-23}
Avogadro's number N_o, molecules/kg-mole	6.02×10^{26}
Permittivity of free space ϵ_o, farad/m	8.854×10^{-12}
Permeability of free space μ_o, henry/m	$4\pi \times 10^{-7}$
Speed of light, m/sec	3.00×10^8
Wavelength range of visible light, A	4000 − 7200
Total blackbody radiation, watt/m^2 °K^4	5.68×10^{-8}

† From E. M. Conwell, *Proc. IRE*, **40** (Nov. 1952), pp. 1327–1337, and *Proc. IRE*, **46** (June 1958), pp. 1281–1300.

‡ All but the last two are from W. Shockley, *Electrons and Holes in Semiconductors*, D. Van Nostrand Co., Inc., New York, 1950.

Indicates spins of valence-shell electrons.

Standard "Strukturbericht" symbol with prefix "A" omitted (see table at lower right). When an element is polymorphic, the phase stable at room temperature is underlined. • means undetermined; + indicates several phases; L indicates liquid; G indicates gas.

	I	II
	1↑	1↑ 1↓
1s	1 H Hex, $\underline{G}$	
2s	3 Li 3, $\underline{2}$, 1	4 Be 3
3s	11 Na 3, $\underline{2}$	12 Mg 3
4s	19 K 2	20 Ca $\underline{1}$, 3, 2
5s	37 Rb 2	38 Sr $\underline{1}$, 3, 2
6s	55 Cs 2	56 Ba 2
7s	87 Fr •	88 Ra •

4f	57–70 See * below
5f	89–102 See † below

	III	IV	V	VI	VII	VIII			I	II
	1↑	2↑	3↑	4↑	5↑	5↑ 1↓	5↑ 2↓	5↑ 3↓	5↑ 4↓	5↑ 5↓
3d	21 Sc 3	22 Ti $\underline{3}$, 2	23 V 2	24 Cr 2	25 Mn $\underline{12}$, 13, 1, 2	26 Fe $\underline{2}$, 1, 2	27 Co $\underline{3}$, 1	28 Ni 1	29 Cu 1	30 Zn 3
4d	39 Y 3	40 Zr $\underline{3}$, 2	41 Nb 2	42 Mo 2	43 Tc 3	44 Ru 3	45 Rh 1	46 Pd 1	47 Ag 1	48 Cd 3
5d	71 Lu 3	72 Hf $\underline{3}$, 2	73 Ta 2	74 W 2	75 Re 3	76 Os 3	77 Ir 1	78 Pt 1	79 Au 1	80 Hg 10, $\underline{L}$

	III	IV	V	VI	VII	VIII
						1↑ 1↓
1s						2 He 3, $\underline{G}$
	1↑	2↑	3↑	3↑ 1↓	3↑ 2↓	3↑ 3↓
2p	5 B Tetr. +	6 C 4, 9, +	7 N +, $\underline{G}$	8 O +, $\underline{G}$	9 F •, $\underline{G}$	10 Ne 1, $\underline{G}$
3p	13 Al 1	14 Si 4	15 P 16+	16 S 16+	17 Cl •, $\underline{G}$	18 A 1, $\underline{G}$
4p	31 Ga 11	32 Ge 4	33 As 7	34 Se 8, +	35 Br 11, $\underline{L}$	36 Kr 1, $\underline{G}$
5p	49 In 6	50 Sn 4, $\underline{5}$	51 Sb 7	52 Te 8	53 I 11	54 Xe 1, $\underline{G}$
6p	81 Ti $\underline{3}$, 2	82 Pb 1	83 Bi 7	84 Po +	85 At •	86 Rn 1, $\underline{G}$

		1↑	2↑	3↑	4↑	5↑	6↑	7↑	7↑ 1↓	7↑ 2↓	7↑ 3↓	7↑ 4↓	7↑ 5↓	7↑ 6↓	7↑ 7↓
*	4f	57 La $\underline{3}$, 1	58 Ce $\underline{1}$, 3	59 Pr $\underline{3}$, 1	60 Nd 3	61 Pm •	62 Sm Rhb.	63 Eu 2	64 Gd 3	65 Tb 3	66 Dy 3	67 Ho 3	68 Er 3	69 Tm 3	70 Yb 1
†	5f	89 Ac $\underline{1}$	90 Th $\underline{1}$, 2	91 Pa Tetr.	92 U $\underline{20}$, +	93 Np +	94 Pu +	95 Am •	96 Cm •	97 Bk •	98 Cf •	99 Es •	100 Fm •	101 Md •	102 No •

"Strukturbericht" Symbols			
Sym	System	Sym	System
A1	Cubic (F.C.)	A9	Hexagonal
A2	Cubic (B.C.)	A10	Rhombic
A3	Hexagonal	A11	Orthorhombic
A4	Cubic	A12	Cubic
A5	Tetragonal	A13	Cubic
A6	Tetragonal	A16	Orthorhombic
A7	Rhombic	A20	Orthorhombic
A8	Hexagonal		

Fig. 1.3. Periodic table of the elements, showing configuration of outer electrons and crystal structures. (Table prepared at Westinghouse Research Laboratories, Pittsburgh, Pa., by A. J. Cornish.)

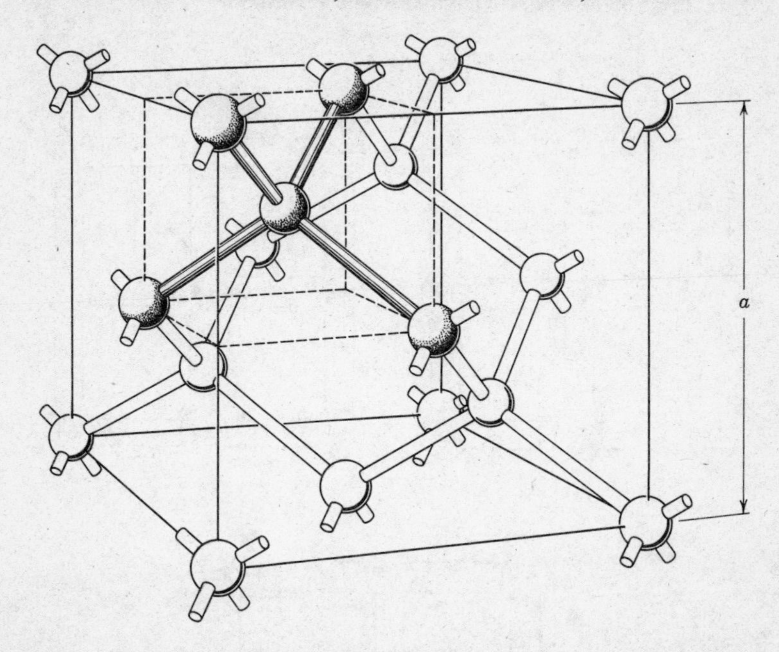

Fig. 1.4. Diamond crystal structure. For C, Si, and Ge: a = 3.56, 5.43 and 5.66 A, respectively. Nearest neighbor spacing is $a\sqrt{3}/4$, and occurs with tetragonal symmetry (dotted cube). The large cube (volume a^3) effectively contains 8 atoms. (After W. Shockley, *Electrons and Holes in Semiconductors*, D. Van Nostrand, Inc., New York, 1950.)

holds together the atoms of the diamond crystal structure. These bonds also dominate the electrical behavior of the crystal. We shall find, in fact, that the valence-bond feature of the diamond structure of germanium and silicon is almost entirely responsible for the semiconducting character of these materials, whereas the high-conducting character of common metals comes from their "electron sea" structure.

Figure 1.5*a* is a simple two-dimensional schematic picture which overemphasizes the covalent bonds. This figure is supposed to convey the idea of a perfectly pure crystal at very low temperatures, when all valence electrons remain bound in the covalent bonds and the material is an insulator. Not only is the whole crystal electrically neutral, but in these circumstances so is each atom (in the dotted circle).

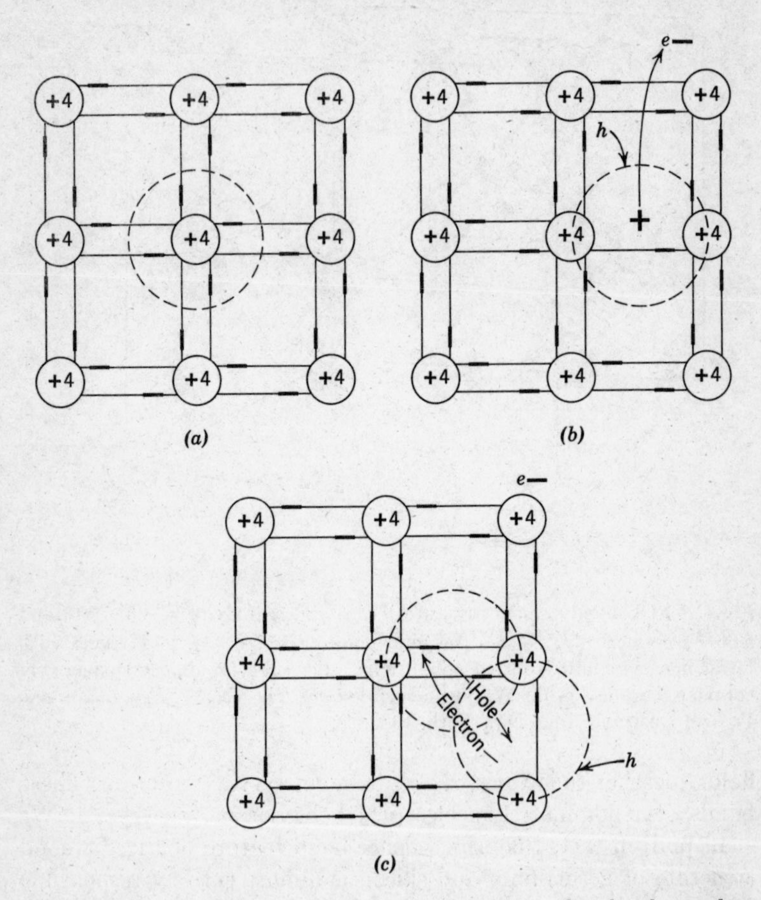

Fig. 1.5. Schematic diamond structure, showing (*a*) covalent bonds at a low temperature (no conduction), (*b*) an overlocalized model for generation of hole *h* and conduction electron *e*, and (*c*) highly oversimplified representation of one possible thermal motion of a hole *h*.

At higher temperatures, however, the thermal vibration of the atoms in the crystal lattice succeeds in shaking loose some of the "bound" electrons, as shown by electron *e* in Fig. 1.5*b*. The energy E_g required to break such a bond is essentially an ionization energy; but because many atoms in the periodic crystal influence the motion of each bound electron, this energy differs in magnitude from ionization energies of isolated atoms. It is convenient to specify energies like these in units of *electron volts* (ev), one of which is the energy gained by an electron in falling through a potential difference of 1 volt. Thus 1 ev = 1.60×10^{-19} joule. For example, in these units the ionization energy is about 0.7 ev in Ge and 1.1 ev in Si, compared to about 8 ev for isolated atoms. When ionization takes place within the solid, the whole crystal still remains electrically neutral, of course, although some of the individual atoms do not. Even at room temperature the number of free electrons is not nearly as large as it would be in a metal; as we shall see, one electron is set free for approximately every 10^9 atoms in germanium, whereas in silicon the corresponding number of free electrons is about 1000 times smaller.

We are not to conclude, however, that the situation is thus made identical with that of a poorly conductive metal. The reasons for the differences are of a most fundamental nature, intimately associated with the real meaning of a covalent bond. Moreover, diode and transistor behaviors depend critically upon these differences. The details of the matter eventually rest on the modern theory of solids, for which quantum mechanics has thus far furnished the only satisfactory basis. Rather than attempt to cover much of the theory in detail here, we present first some workable, oversimplified pictures of the situation in familiar terms, and then try to point out as clearly as we can which features of these pictures can withstand closer examination only by direct recourse to quantum mechanics. In this way we hope to make it easy to go on with the subject matter, but at the same time show as precisely as possible where the reader may some day wish to fill in the argument by further study.

1.3.2 *Conduction Electrons and Holes*

The most startling property which makes a semiconductor so different from a metal is that *the vacancy h (Fig. 1.5b), left in the covalent bond by the liberated electron e, behaves as if it were a new free particle with a positive electronic charge* $+q(= +1.60 \times 10^{-19}$ *coulomb) and a mass comparable to that of an electron. This "apparent" particle is quite logically called a "hole."*

That the atomic neighborhood from which electron e came now has a charge $+q$ (indicated by + in the dashed circle of Fig. 1.5b)

is no surprise; it has been ionized by thermal vibration. The surprise is that this positively charged vacancy can *move*, even though the ionized atom itself does *not* move. One way of visualizing such motion is shown in Fig. 1.5*c* where an electron from a covalent bond in the lower right-hand corner has, as a result of its thermal motion, jumped into the original vacancy shown in Fig. 1.5*b*. The result is that this vacancy, with its associated positive charge unbalance, has apparently moved to the lower right-hand corner of Fig. 1.5*c*. Thus the motion of the "hole" may be looked upon as a *transfer of ionization* from one atom to another, carried out (curiously enough) by the motion of the "bound" electrons between their covalent bonds. Note that the electron originally set "free" by thermal vibration is not needed in this process, and may go its own way in a completely *independent* fashion. It becomes a *conduction electron*.

As important and helpful as the foregoing picture of a hole turns out to be, it is worthwhile to discuss for a moment certain serious limitations of the viewpoint implied by Fig. 1.5. That the motion of a hole is simply a convenient way of describing the actual motions of "bound" electrons, whenever there is an empty bond in the crystal, is adequately and truthfully portrayed by the figure. However, the detailed mechanism through which this is accomplished is not as simple as it may seem from the illustrations.

A rough calculation, based on the Heisenberg uncertainty principle, reveals one reason why Fig. 1.5, showing thermal motion, is oversimplified. Thus, even if we accept the statement that the hole behaves essentially like a free particle with the electronic charge and a mass of the order of that of an electron, we can show that the extent to which this "particle" is localized according to quantum mechanics, *given that it shares random thermal motion with its environment*, is much less precise than what is implied in Fig. 1.5.† In Fig. 1.5, in accordance with the discussion associated with Fig. 1.0, the hole is being localized to within a distance of a very few angstrom units, essentially within the dimension of a single atom. But at room temperature, corresponding to $T = 300°K$, the particle (treated as an ideal-gas particle) would be

† If the hole does not have random thermal motion, its localization will be determined by the precise experimental circumstances used to observe it. Such circumstances are hard to envision, and not of interest here.

expected to have on the average approximately $kT/2$ joules of thermal energy per degree of freedom, where k is Boltzmann's constant. Therefore, if we account for three degrees of freedom, the thermal velocity would be given by the relation

$$\tfrac{1}{2}mv_{th}^2 = \frac{3}{2}kT \qquad (1.1)$$

which at room temperature leads to a velocity $v_{th} = 1.2 \times 10^5$ m/sec, or approximately 10^7 cm/sec.†

If we argue for simplicity that the random thermal motion of a particle places its energy with about equal likelihood anywhere in the range between about one-third and three times its mean value, the particle momentum would correspondingly be uncertain by about its whole mean value. Thus $\Delta p \approx mv_{th}$, and the corresponding uncertainty in position Δr is given by the Heisenberg principle as

$$\Delta r \approx \frac{h}{\Delta p} \approx \frac{h}{mv_{th}} \qquad (1.2)$$

in which h is Planck's constant. In the present case, then, $\Delta r = 60\text{A}$.

Inasmuch as the atomic spacing in the crystal lattice is, according to Fig. 1.4, approximately 2 to 2.5A between nearest neighbors, a cube about 2.5A on each side may be assigned to each atom. It follows that the localization of the hole within a sphere of diameter $\Delta r = 60\text{A}$ would have it spanning about 7000 crystal atoms. Clearly, any notion that a hole in its thermal wanderings jumps between neighboring bonds is without physical significance!

It is also extremely important to recognize that the same calculations made in Eqs. 1.1 and 1.2 apply approximately to the "size" of a "conduction electron" in the crystal. They indicate that it too is greatly overlocalized in pictures of the type shown in Fig. 1.5. However, it is easy to prove that the atoms of the lattice are quite well-localized because of their large masses (P1.4).

When considering dynamics, it is hard to avoid the implication in Fig. 1.5*c* that the motion of a hole is always "really" produced

† In the dynamical laws like Eq. 1.1, it is important to use a self-consistent set of units, unless conversion factors are added. Thus the relation as written is correct in the mks system, but *not* in a mixed system with cm in place of m.

by an oppositely directed motion of *one particular* electron from a valence bond lying at least somewhere along the eventual path of the hole. We might even be tempted some time later to imagine precisely such a motion in the presence of an electric field, pushing a valence electron to the left, to create a motion of a hole to the right. Unfortunately, such reasoning is very treacherous and can easily lead us into errors; an example occurs in the deflection of a current of moving holes by a magnetic field (see Sec. 1.5.3 and Laboratory A2.4 on the Hall effect). It is true that our calculations, given above, show that the hole is not localized nearly enough to support the details of such a "retrograde electron motion" point of view; but this fact is easy to gloss over and to forget. Perhaps it is helpful to think of the way in which a bubble moves through a liquid, in order to get the idea that the "bound" electrons actually act as a *group* to produce the motion of a hole. Pictures like Fig. 1.5 ought, therefore, to be considered only as mnemonics that allow us to keep in mind the principal results of significance concerning the conduction electrons and holes, and not as accurate representations of their dynamics.

It should begin to be clear now that the main difficulty in the present problem is that, because they tend to be localized on a small space scale (of the order of atomic dimensions), *"bound" electrons do not behave like familiar classical electrons at all.* To understand their motion properly, a knowledge of quantum mechanics is essential. The novelty of the situation lies in the very remarkable circumstance that, whenever a valence bond is empty, the *classical* motion of an ordinary positively charged particle can nevertheless correctly *represent* the essentially quantum-mechanical motion of all the actual "bound" electrons. At least this is true of the electromagnetic properties of the semiconductors in which we shall be interested here. Therefore, if we wish to discuss semiconductor electronics in terms of the familiar language of *classical* physics, we *must* adopt the concept of a hole. That is, we must discuss the motion of the hole *instead* of that of the "bound" electrons which are actually doing the moving.

1.3.3 *Effective Mass*

We shall soon have some need to discuss the acceleration of holes (and of conduction electrons, for that matter) which results from the application of new forces to the material. These forces, usually in the form of applied electric and/or magnetic fields, arise for example in connection with a measurement of electrical conductivity (applied electric field) or of Hall effect (applied electric and magnetic fields). At first, it may seem appropriate simply to equate the acceleration times the mass of an electron to the force of the applied field in order to determine the motion, on the theory that, after all, electrons are actually doing the moving in response to the applied forces. But, on second thought, this idea is seen to be incorrect because the electrons in the solid are acted upon not only by any applied electric or magnetic forces in question, but also by electrical forces from various atomic cores and other electrons in the periodic crystal array. As we indicated earlier, the size scale for computing the atomic and electronic forces is so small that quantum mechanics is essential to the determination of the motion produced by them. Accordingly, the *additional* accelerations produced by the added forces cannot be computed by Newton's laws, based only on classical principles.

Indeed, it turns out that a description which even resembles a classical one, either for the behavior of a hole or of a conduction electron in the solid, is valid only if the applied forces are much weaker than those associated with the spatially periodic electric field produced by the lattice atoms. In other words, the potential of the applied forces must not vary much over a few interatomic distances of the crystal structure. Under these circumstances, it does turn out to be possible *to represent the effect of all the periodic crystal forces in a parameter known as the effective mass*, denoted by the symbol m^*. Thus the equation of motion for a hole, insofar as we are concerned with its response to weak applied forces $\mathbf{F}_h$, superimposed on the normal periodic crystalline forces, is given by

$$\mathbf{F}_h = m_h^* \frac{d\mathbf{v}_h}{dt} \tag{1.3}$$

in which the subscript h has been used to refer to the properties of a hole. Similarly, for the "free" or conduction electrons in the solid, an effective mass m_e^* must be defined if the classical equation of motion

$$\mathbf{F}_e = m_e^* \frac{d\mathbf{v}_e}{dt} \tag{1.4}$$

is to be employed to compute the response of the particles to an applied force.

A classical picture of holes and conduction electrons is possible only because the major quantum features of the electronic motions in the solid can be buried in the "effective mass" parameter m^*.

The quantum constraints show through, however, in the curious fact that the values for the effective masses of conduction electrons and holes in various materials may lie in a range from less than $\frac{1}{100}$ to greater than 1 times the normal mass of an isolated electron! The values larger than the normal mass may seem easy to imagine in terms of electrons which are relatively tightly bound. But conduction-electron effective masses *less* than the normal mass, and any particular values of the effective masses of holes, cannot be visualized in classical terms at all. The fact that small effective masses are very common indicates the essentially quantum nature of the problem. As a matter of fact, in many physics problems a careful treatment would require use of different values of effective mass in the three space directions, so the "effective mass" becomes not just a single number but a matrix or tensor. For the cases of semiconductor devices to be considered here, however, it will be sufficient to take a single number for the effective mass of each species of carrier in each semiconductor material, and we shall do so without further comment.

Evidently the safest view is that the effective mass gives a quantum-mechanical measure of the ease with which an external force can accelerate the electrons and holes in the solid through the periodic hurdles supplied by the internal atomic potentials. Note that the effective mass is therefore appropriate *only* to motions of the carriers *with respect to the crystal lattice* (P1.6*a* and *e*).

The effective-mass idea means that *both* the "conduction" electron and the hole are *phenomenological particles* that do not exist outside the solid. It would be pleasant if we could take the time to develop this notion of the effective mass carefully from quantum theory, and then, because of its convenience, use it instead of quantum theory. Unfortunately, the extended argument required to do this is at best not as clear as we might wish, and it certainly involves issues which are far outside our present theme. Our procedure here will instead be to believe in holes, and to accept and use the effective-mass concept for holes and conduction electrons in its full generality, on the basis not only of the oversimplified examples presented in Fig. 1.5, but also of some excellent experiments (see Secs. 1.5.3, 1.6.1, and 4.3.6, as well as the films referred to there and the Laboratories A1–A5). These do render the idea very plausible, and make adequately clear the important fact that holes are merely simpler but equivalent representations for the electromagnetic behavior of "bound" electrons.

At this point we can summarize the consequences of our model of a pure semiconductor as follows:

(1) In semiconductors such as germanium and silicon electrical conduction may take place at ordinary temperatures via *two distinct and independent quantum-mechanical modes of electron motion, which can most simply be described instead as classical conduction by "conduction" electrons* with charge $-q$ and effective mass m_e^*, *and by holes* with charge $+q$ and effective mass m_h^*.

(2) In *very pure* germanium and silicon, conduction electrons and holes always occur in equal numbers, since the process of "freeing" a bound electron must perforce leave a vacancy (hole) behind. Thus *two current carriers are made available every time an electron is set "free,"* and the material thereby exhibits *intrinsic conductivity*.

1.4 THE ROLE OF IMPURITIES: EXTRINSIC CONDUCTION

1.4.1 *Donors*

Because conduction electrons and holes always occur in equal numbers in very pure semiconductors, such material is insufficiently flexible for the most important practical applications of semiconductors. In fact, the development of practical semiconductor devices would not be possible without the freedom of adding minute, controlled amounts (of the order of 1 part per million) of appropriate foreign substances (impurities) to otherwise pure material. In this way it is possible to create semiconductors which exhibit *extrinsic conductivity* over a reasonably wide temperature range. In this range, these semiconductors have either more conduction electrons than holes, or vice versa. Such a result cannot generally be accomplished on a large scale by simply adding electrons to, or removing them from, the original pure material, since this action upsets the electrical neutrality of the whole crystal and brings into play tremendous electrical restoring forces. We can, however, substitute for a few of the original semiconductor atoms some different ones which have either one more or one less valence electron, as shown schematically in Fig. 1.6.

If we choose an atom which has one more outer electron (i.e., has valence 5 and comes from group V of the periodic table) but is otherwise roughly the same as the original, it will fit into the valence-bond structure without much difficulty (Fig. 1.6*a*), except that when it is electrically neutral (as shown by the dashed circle) its extra electron is like the proverbial "fifth wheel." Because this electron is not really functioning in a valence bond to hold the crystal together, the ionization energy required to set free this extra electron is very small compared to that involved in breaking a covalent bond. As a result, even at quite low temperatures this electron (*e* in Fig. 1.6*a*) is shaken free, as shown, to become a conduction electron. It is most important to observe, however, that the positive ion left behind (inside the dashed circle) has exactly the same number of electrons around it as does any of the original semiconductor atoms. There is no vacancy *in the bond structure*

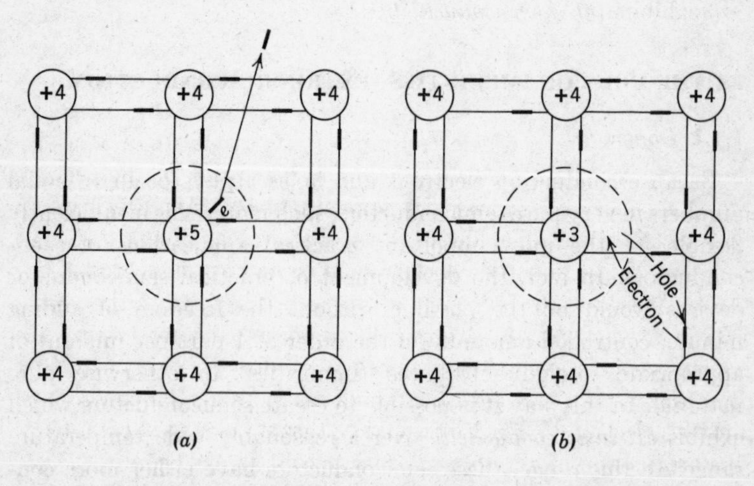

Fig. 1.6. Schematic diamond structure, showing (*a*) donor atom and (*b*) acceptor atom. The electrons and holes are again greatly overlocalized.

into which other bound electrons can slide. The positive ion is therefore a *fixed* charge, which *cannot* help to carry current; in short, it is *not* a hole. It need hardly be added that the crystal as a whole remains electrically neutral, whether or not electron e is free.

Besides the difficulty with overlocalization of holes and free electrons which we have mentioned before, there is another problem with scale in Fig. 1.6. The electron bound to the foreign atom in Fig. 1.6a cannot be as close to the effective nucleus as shown in the drawing; otherwise the energy required to free it would be at least comparable to that needed to break a valence bond. The extra electron would not come free any more easily than the rest. Let us examine the possibility that it is much farther away, so it includes within its orbit a large number of the normal crystal atoms, and find out whether the increased distance is consistent with the observed fact that the energy needed to "free" this extra electron is small.

The significant point of the issue at hand is the electric polarization of the crystal lattice. Upon the application of an electric field, the nuclear charges of the atoms in the crystal lattice, and the charges of the relatively tightly bound electrons which do *not* take part in the normal conduction process, undergo a distortion in relative position which yields electric polarization. The particular electr ons involved can displace themselves slightly under the action of the applied field, even though they cannot actually break away into free motion. The resulting electric polarization is normally described by the dielectric constant of the material. In semiconductors such as silicon and germanium the value of the relative dielectric constant is quite high, namely, about 16 for germanium and 12 for silicon (Table 1.0). Such high values of the dielectric constant imply that the effective field of any atomic core is very much reduced by the presence of the polarization from the neighboring normal crystal atoms whenever the distance from the nucleus includes even a relatively small number of these neighbors. Accordingly, the first guess we would make about the behavior of the fifth electron in Fig. 1.6a would be that it behaves substantially like an electron in a hydrogen atom, except for two modifications: first, the appropriate dielectric constant for the computation of the electric field of the effective nucleus would be that of the semiconductor material rather than that of free space; second, the appropriate mass of the electron would be that relevant to its motion through the periodic crystal lattice rather than that appropriate to free space.

On this basis the Bohr model for a hydrogen-like atom may be used to make a rough first estimate of the radius and energy of the lowest state, as follows. First, the balance of the electric and centrifugal forces of a circular orbit requires

$$\frac{q^2}{4\pi\epsilon r^2} = \frac{m_e^* v^2}{r} \tag{1.5}$$

or

$$\frac{m_e^* v^2}{2} = \frac{1}{2}\frac{q^2}{4\pi\epsilon r} \tag{1.6}$$

which indicates that the kinetic energy in the orbit is equal to one-half of the magnitude of the potential energy. Second, the deBroglie postulate assigns to the electron a wavelength λ given by

$$\lambda = \frac{h}{p} = \frac{h}{m_e^* v} \tag{1.7}$$

and the Bohr condition for the lowest orbit requires

$$\lambda = 2\pi r \tag{1.8}$$

From Eqs. 1.7 and 1.8 we may find one relation between the orbital radius and the orbital speed,

$$r = \frac{h}{2\pi m_e^* v} \quad \text{or} \quad v^2 = \frac{h^2}{4\pi^2 m_e^{*2} r^2} \tag{1.9}$$

Finally, elimination of the velocity from Eqs. 1.6 and 1.9 yields for the smallest Bohr radius the result:

$$r_o = \frac{\epsilon h^2}{q^2 m_e^* \pi} = \left(\frac{m}{m_e^*}\right)\left(\frac{\epsilon}{\epsilon_o}\right) r_{o\ \text{Hydrogen}} \tag{1.10}$$

The total energy in the orbit is a sum of the kinetic and potential energies. But we must take note of the fact that the potential energy is actually proportional to the negative of the right-hand side of Eq. 1.6, because the zero of potential is being taken at infinity and we are considering the potential energy of a negative charge in the vicinity of a positive nucleus. Thus, in view of Eq. 1.6, we find for the total energy:

$$E_o = \frac{m_e^* v^2}{2} - \frac{q^2}{4\pi\epsilon r_o} = -\frac{q^2}{8\pi\epsilon r_o} \tag{1.11}$$

which, in terms of Eq. 1.10, becomes

$$\begin{aligned} E_o &= -\frac{q^4 m_e^*}{8h^2\epsilon^2} = \left(\frac{m_e^*}{m}\right)\left(\frac{\epsilon_o}{\epsilon}\right)^2 E_{o\ \text{Hydrogen}} \\ &= -13.6\left(\frac{m_e^*}{m}\right)\left(\frac{\epsilon_o}{\epsilon}\right)^2 \text{ev} \end{aligned} \tag{1.12}$$

The number 13.6 ev comes from the ionization potential of hydrogen gas (P1.5).

Accordingly, if we use a representative value of $0.6m$ for the effective mass of conduction electrons in germanium, and use the relative dielectric constant of 16 (Table 1.0), we obtain a lowest energy for the fifth electron on the impurity atom of about 0.03 ev. This, then, becomes the estimated "ionization" energy for the impurity atom when it is in the crystal, because the zero of energy in Eq. 1.11 was taken for a particle at rest at infinity. In silicon, for group V impurities, we make use of the effective mass m_e^* of approximately m, and the dielectric constant of 12, which leads to an ionization energy for the impurity of about 0.1 ev. Notice that the only characteristic of the impurity atom that appears in the result is its difference in charge from the lattice atom it replaces. This fact suggests that the ionization energy is the same for all impurities that have the same charge.

It is wise before proceeding further to check on the radius of the orbit implied by Eq. 1.10, to see whether it is large enough to justify the approximation involving the use of the dielectric constant. This justification requires the inclusion within the orbit of a large number of ordinary lattice atoms. The value of the first orbital radius r_o for hydrogen is 0.53A (P1.5); so, in view of the effective mass and dielectric constant values required in Eq. 1.10, we find a radius r_o in germanium approximately 30 times that in hydrogen, or 16A. There are about 1000 atoms inside a sphere of this radius, which is a number sufficient to justify the required

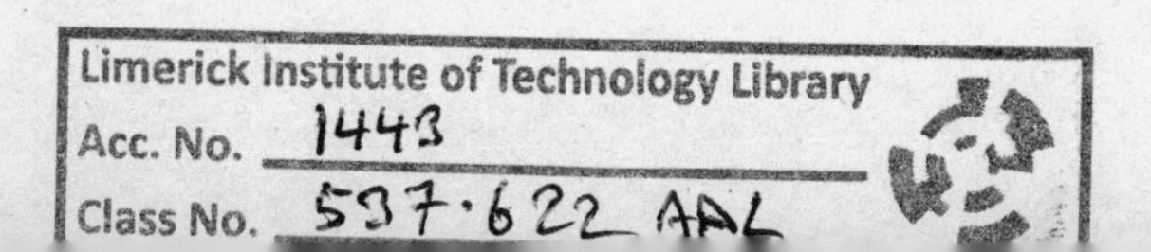

approximations and to make the details of the impurity atom, except its charge, unimportant to the extra electron. For silicon, however, the value of the radius would be only 12 times that of hydrogen, or about 6A, which is small enough to suggest much more doubt about the result than is the case for germanium. If we look ahead to Table 1.1 (p. 25), however, we find the ionization energies to be somewhat smaller than our rough calculations indicate, and, correspondingly, the orbit radii must be somewhat larger than the values we have found. But there is no doubt whatsoever about the order of magnitude of the ionization energy.

The ability of appropriate elements of valence 5 from group V of the periodic table to *donate* a free electron to the crystal *without at the same time producing a hole*, has earned for them the name *donors;* they are said to render the semiconductor n-type because it then carries current mostly by *negatively* charged conduction electrons. We must hasten to add, however, that there are always at least a few holes at any temperature above absolute zero, because at least a few ordinary covalent bonds are always being broken thermally. We cannot afford to forget about these, in spite of their relatively small number at room temperature in n-type material. For this reason we speak of the conduction electrons as the *majority* (not exclusive) *carriers* in n-type material.

1.4.2 *Acceptors*

The situation opposite to that just described in the last section is realized by substituting for an original atom a foreign element of similar size which has one less outer electron (i.e., has valence 3 and comes from group III of the periodic table), as illustrated in Fig. 1.6*b*. When neutral, this atom fails to fill up the bond structure in its neighborhood (the dashed circle). Once again, we must keep in mind that the bonding forces in a valence crystal are very strong; they form the major agent that holds the atoms together in the crystalline array. Therefore, the vacancy shown in Fig. 1.6*b* is almost as attractive to the ordinary bound electrons as it would be if it occurred near one of the original crystal atoms. The only difference is that the latter would be positively charged, whereas the foreign atom is neutral. But the electrostatic forces on the electrons in the crystals we are considering are much smaller than the major bonding forces, as we have implied before. Consequently, even at relatively low temperatures and without the help of any electrostatic attraction, the other bound electrons are easily persuaded to slide into this vacancy, as shown in the figure, so that a *hole* is made available for conduction. Of course, the foreign atom becomes negatively charged (negatively ionized) when its valence bond is filled; but since the atom cannot move, *it cannot contribute any negative carrier to the conductivity.*

Actually the remarks previously made about the scale of Fig. 1.6*a* apply also to the scale of Fig. 1.6*b*. The easiest way to understand this fact is to regard the situation in Fig. 1.6*b* in a slightly different light than the immediately preceding description would suggest. In accordance with the idea that the hole is also described as an ordinary classical particle with an effective mass m_h^* and a charge $+q$, in just the same way that the electron has been similarly described, it is reasonable to look at Fig. 1.6*b* as if, in the normal condition, the acceptor atom had a hole orbiting about it, just as the electron was orbiting about the donor in Fig. 1.6*a*! That is, in the present case if the hole were removed by filling it with an electron, the effective core would be negatively charged by the amount $-q$. The hole may then be regarded as added to neutralize the structure, and the orbit of this hole (which is much larger than is suggested by the picture) can be computed by the same means used in Eqs. 1.5 through 1.12. The only difference would be the replacement of the effective mass of the electron by that of the hole. The calculation must be done by using the effective mass of a hole in the appropriate material. According to Table 1.0 and Eq. 1.12., the binding energy of the hole to the acceptor is of the order of 0.02 ev for germanium and 0.06 ev for silicon. For the acceptors, Table 1.1 again shows somewhat smaller ionization energies than our rough calculations yield, but it also definitely shows more variation of binding energy with specific acceptor atom in Si than in Ge. This agrees with our picture (Eq. 1.10) predicting a smaller impurity orbit radius in Si than in Ge, because of the smaller ϵ and larger m^* in Si.

Atoms of appropriate size from group III of the periodic table, and with valence 3, are known as *acceptors* because of their willingness to *accept* bound electrons. Because they add holes to the material without adding corresponding conduction electrons, they are said to render the material *p-type*, inasmuch as it now conducts mainly, but not quite exclusively, by *positively charged holes.*

Holes, then, are the *majority carriers* in *p*-type material, and the *minority carriers* in *n*-type material. Similarly, conduction electrons are the majority carriers in *n*-type material and the minority carriers in *p*-type material. These matters are summarized in Table 1.1 for several donor and acceptor impurities that are commonly added to germanium and silicon.

1.5 THE CONDUCTION PROCESS

1.5.0 *Introduction*

From our point of view, the most important thing about semiconductors is their ability to carry electric current. Inasmuch as current is defined as the time-rate at which charge is transported across a given surface in a direction normal to it, the current will depend on both the number of charges free to move and the speeds at which they move. It is important to remember that the currents we generally measure are the results of the motions of a great many individual charges and not the "component" current produced by the motion of any one.

We have already pointed out that in metals the usual situation is one in which the number n of conduction electrons per cubic centimeter is large and virtually independent of temperature over a very wide range. Of course, both the crystal atoms and the conduction electrons are always in random thermal motion; the atoms vibrate about their mean positions in the crystal lattice, and the electrons fly very rapidly about in all directions, making numerous random "collisions" with the vibrating atoms. Under these conditions, the electrons as a group carry no time-average current in any direction, because on the average as many are moving in one direction as in any other. There is nevertheless a measurable alternating component of current associated with this random motion, which accounts for the thermal-agitation noise (Johnson noise) in resistors.

Now let us be sure we visualize what is going on between the electrons and the lattice in a little more detail. We have agreed to think of the current carriers as particles with a certain effective mass, and to describe the electronic behavior of the solid in these

TABLE 1.1

Some properties of silicon and germanium, and impurities commonly used to control their electrical conductivities.

Element	Number of Valence Electrons	Atomic Number	Radius in Crystal (Angstroms)	Function	Majority Free Carrier	Electron Volts Ionization Energy to Produce Free Carrier		Conductivity Type Created
						in Ge†	in Si‡ †	
Boron (B)	3	5	0.81	Acceptor	Hole	0.0104	0.045	*p*-type
Aluminum (Al)	3	13	1.26	Acceptor	Hole	0.0102	0.057	*p*-type
Gallium (Ga)	3	31	1.26	Acceptor	Hole	0.0108	0.065	*p*-type
Indium (In)	3	49	1.44	Acceptor	Hole	0.0112	0.16	*p*-type
Tin (Sn)	4	50	1.40	Neutral Solder	. . .	. . .	. . .	. . .
Phosphorus (P)	5	15	1.13	Donor	Electron	0.0120	0.044	*n*-type
Arsenic (As)	5	33	1.18	Donor	Electron	0.0127	0.049	*n*-type
Antimony (Sb)	5	51	1.36	Donor	Electron	0.0096	0.039	*n*-type
Silicon (Si)	4	14	1.17	Host	Equal	$1.205 - 2.8 \times 10^{-4} T(°K)$		intrinsic
Germanium (Ge)	4	32	1.22	Host	Equal	$0.782 - 3.9 \times 10^{-4} T(°K)$		intrinsic

† R. A. Smith, *Semiconductors*, Cambridge University Press, 1959.

‡ E. M. Conwell, *Proc. IRE*, **46** (June 1958), pp. 1281–1300.

terms. This very remarkable point of view "explains" at once why it would not be necessary for us to consider any collisions of the electrons with, or "scattering" of them by, the crystal atoms if these atoms were sitting in their equilibrium positions in a *perfect* crystal. It will be recalled that all the effects of the *periodic* lattice potentials in the crystal are buried entirely in the effective mass parameter m^*, so that only *departures* from periodicity act as deflecting forces. Moreover, to use the effective-mass idea in treating the deflection of carriers by these irregularities, the irregularities should occur over distances that are large compared to a lattice spacing. It is here worth discussing four types of such disturbances which act to deflect the conduction electrons or holes in their wanderings through the usual lattice.

At normal temperatures, by far the most important deflecting or "scattering" events arise from distortions of the crystal associated with thermal vibrations. Atoms are either squeezed together over a region, or pulled apart over a region, as compared to their normal spacing in the ideal periodic lattice. These displacements obviously alter the electric potential in their neighborhoods, and these potential differences correspond to "applied" *electric fields* which alter the direction and magnitude of the particle's momentum. Such events are called *thermal* or *lattice scattering*.

To establish the important size scale of the disturbances described above, we must remember that the conduction electron and hole are perceived by the lattice as being rather large at room temperature, with effective diameters of the order of 10 lattice spacings. On this fairly big scale, we can think of the lattice as being almost continuous in structure and vibrating in various wave patterns like a stretched string clamped at its ends. The compressions and rarefactions of the "standing waves" of vibration are the regions that deflect the carriers. At a given temperature many such standing waves of different wavelengths are excited thermally, with peak compressional energies proportional to the thermal energy kT (or compressional amplitudes proportional to $\sqrt{kT}$).

Now, for a given peak amplitude, the most effective distortion regions for deflection purposes are those compressions or rarefactions which extend over approximately the full "effective size" of the carrier. That is, a standing wave whose alternate compressions and rarefactions (each $\frac{1}{2}\lambda$ in size) occupy a "diameter" of the carrier should dominate in scattering it. Waves with much shorter wavelengths have several rarefactions and condensations within the carrier diameter, so the effects of successive condensations and rarefactions cancel out each other. Waves with much longer wavelengths impose over the carrier body only weak electric forces (the potential gradient is less), and are therefore also ineffective for scattering. Accordingly, the standing wave with $\lambda/2 \approx \Delta r$ (see Eq. 1.2 with m^* instead of m) is of greatest interest.

Of course, these compressional standing waves are essentially sound waves, and the speed of sound in most of the materials we are considering is of the order of 2×10^5 cm/sec, whereas the thermal speed of the carriers has been shown to be about 10^7 cm/sec. Thus the carrier moves 50 wavelengths while a sound wave moves one wavelength; or the carrier moves through 50 compressions and 50 rarefactions in one period of the standing wave. Naturally, in one period the standing wave has changed a lot—but in $\frac{1}{5}$ period, for example, it has not changed much. Thus, on the average, a carrier moves through perhaps 20 significant "bumps" in the potential before the pattern of the bumps changes appreciably. It is, accordingly, quite clear that the random motion of the carriers actually consists of a complicated warping of paths in a relatively gradual manner, in which the individual events need not be very violent, but the cumulative effects may well be large.

The second most important type of scattering factor in the lattice is the charged impurity atom; for example, an ionized donor or acceptor. We have seen that even for an electron *bound to* (that is, orbiting about) such an impurity, the radius is quite a few lattice spacings; so it is clear that appreciable *deflection* of a moving charge carrier will occur at distances much greater than this value. Besides the thermal or lattice scattering, this *impurity scattering* may become important, depending on the temperature and the impurity concentration. Of course, the two forms may be combined if a vibration compression occurs in a region of the crystal containing a charged impurity atom in it. The two processes are mixed up in space, but usually act independently on the carriers because there is no reason for them to occur with any correlation to each other (P1.9). The charged impurities of either sign will deflect carriers of either sign, by attraction or repulsion as appropriate, and with an effectiveness which depends on the initial aim of the moving particle with respect to the fixed one.

Third, even an *electrically neutral* impurity atom can scatter a carrier, either when the carrier gets close enough to the impurity to detect its positive core separately from its orbiting outer electrons, or when the impurity atom produces a local lattice strain and a corresponding bump of electric potential. This kind of scattering is weak and usually obscured by the other mechanisms.

Finally, we must consider the possibility of conduction electrons and holes deflecting each other in passing. The situation is similar to that of scattering by an oppositely charged impurity, except that both particles deflect about their common center of mass. This type of scattering is not likely to be important for the *majority* carrier unless *both* types of carriers are in plentiful supply, and in amounts large enough to compete with the irregularities from thermal vibration and the presence of charged donors and acceptors. These conditions are not common, as we shall understand better shortly; so scattering of holes by conduction electrons, or vice versa, is rarely of major importance in electrical conduction.

We should also concern ourselves with possible deflections produced by encounters between two conduction electrons or two holes. But we must remember that we are interested in the total current carried by a large group of the carriers. Take positive carriers, for example. The current is the sum of the individual currents carried by each particle.

$$\mathbf{J} = \sum_j q\mathbf{v}_j = \frac{q}{m^*} \sum_j \mathbf{p}_j = \frac{q}{m^*} \mathbf{P}$$

Thus the current is proportional to the momentum of the *entire group* for carriers of a single type, and in collisions between these same carriers the *total* momentum is conserved. Whereas the path of any individual carrier is much altered by collisions with others of its own type, the net effect upon the measured current from all the carriers of that type is zero for this scattering process. Therefore, we need not here consider further the collisions of conduction electrons with each other or of holes with each other in our thinking about the current.†

† In the vast majority of cases, this conclusion is in accord with experiment. But our derivation of the result from elementary arguments is greatly oversimplified.

The foregoing discussion concerned the behavior of the carriers in the absence of any field. There are large individual currents, but they cancel out on the average over the group. When an electric field is applied, however, *all* the carriers of a given type are acted upon by it in the *same* way. There is a steady tendency for them all to be pushed in the direction of the field force. Note that to get a very substantial component of current in the direction of the field force from a great many carriers, very little *organized* velocity in that direction is needed from each one.

Now, although the field accelerates each carrier, it only slightly increases the carrier velocity. The individual circumstances of each deflection discussed above are altered very little. These deflections are controlled by the much greater thermal motions of the carriers and the lattice atoms involved. Hence, the very first effect of the increased speed is to make the deflections come *more often*, at about the same strength. The key point is that those features with which the carriers collide in the lattice, and which affect their *total* momentum, tend to be laid out with an average *space* pattern reminiscent of a pin-ball machine. Consequently, the faster the carrier moves, the more collisions it makes *per unit of time*. With the nature of each collision substantially unaltered, the resulting force increases with speed, as viewed in an average over the whole group of carriers. This is the same kind of opposition supplied by viscous friction, and we know that such friction leads to *terminal velocities* in free fall through the atmosphere, etc. Thus we expect the field, if it is not too large, to do the same thing to the carriers; i.e., to produce a *drift speed* rather than a steady acceleration. Of course, the latter would be impossible anyway, because it would prohibit a steady state and cause an ever-increasing current (along with an ever-increasing power input)!

It is important to realize that it is not primarily the loss of *energy* in each collision that limits the current, but the random changes in momentum *direction* for the members of the group that limit it. As can be seen from our description of the scattering processes that are significant in connection with the current, the energy loss *per collision* is very small, essentially because the carrier mass is small compared to the mass of the crystal atoms (P1.7). However, the energy loss in the whole group of carriers *per unit time* can easily be important, because of the exceedingly large number of collisions involved. Thus the Joule heating associated with current flow through a resistance is far from negligible, and is attributable entirely to the collisions which establish the drift velocity of the carriers, even though the energy loss per event is very small indeed compared to random thermal energy.

1.5.1 *Mobility*

We have pointed out that insofar as *additional* carrier velocities are concerned, over and above the thermal ones, the effect of "collisions" in resisting a weak applied force is similar to that of viscous friction. But both the frequency and severity of collisions in equilibrium is different for carriers with different values of thermal energy. We are obliged to keep in mind that there may really be a significant spread of particle *thermal* energies and speeds at any given temperature, even though the incremental energy and speed change produced by the applied force is very small.

Therefore, the small changes in number of collisions per unit time produced by the velocity increments resulting from the applied force will generally be *different* for groups of carriers whose total kinetic energies are different. We should accordingly find the response (current, for instance) to any given applied force *for each such group separately*, and then add up the responses for all the groups. To carry out this procedure in detail requires that we know how the carriers are distributed in energy under nonequilibrium conditions, so that we will know the numbers of carriers that have kinetic energies in any arbitrary small range between E and $E + dE$, for example. In fact, we shall take up some similar questions in Chapter 3, but only for thermal equilibrium conditions. Fortunately, however, there is much we can do simply by understanding that a nonequilibrium distribution must be accounted for, without knowing its form in detail.

To carry out the analysis explicitly, denote by $\mathbf{v}_i$ the *average incremental* velocity imparted by an applied force $\mathbf{F}$ *per particle* to the particles in a group $\delta n(E)$, all of whose energies are in the small range between E and $E + dE$. This group is small enough to permit dE to be small, *but large enough so that random effects cancel out.* As a result of the increased number of collisions per unit time, the incremental velocity gives rise to an effective "resisting force" on the group proportional to $\mathbf{v}_i$; say, $\alpha(E)\mathbf{v}_i\delta n(E)$, where α may depend on the kinetic energy E for the reasons discussed above. Thus the dynamical equation of motion relating the applied force and the incremental velocity *of the whole group* is

$$\mathbf{F}\delta n(E) = m^*\delta n(E) \frac{d\mathbf{v}_i}{dt} + \alpha(E)\mathbf{v}_i\delta n(E) \qquad (1.13a)$$

Note that $\mathbf{F}$ includes only forces over and above the periodic forces of the crystal lattice, these latter being included already in the effective mass m^*. Also, all random velocities cancel out over the group $\delta n(E)$; that is why the equation of motion relates only to the incremental velocity $\mathbf{v}_i$ of the group.

Dividing through Eq. 1.13*a* by the factor $m^*\delta n(E)$ gives

$$\frac{\mathbf{F}}{m^*} = \frac{d\mathbf{v}_i}{dt} + \frac{\mathbf{v}_i}{\tau_c(E)} \qquad (1.13b)$$

in which $\alpha(E)/m^* \equiv 1/\tau_c(E)$. We can show that $\tau_c(E)$ has the significance of a "relaxation time" for the energy group $\delta n(E)$. That is, suppose no force is applied ($\mathbf{F} = 0$), but at $t = 0$ there was an initial disturbance $\mathbf{v}_i(0)$ of the group from its random equilibrium state. Then Eq. 1.13*b* can be solved to yield

$$\mathbf{v}_i = \mathbf{v}_i(0)e^{-t/\tau_c(E)}; \qquad t \geq 0 \qquad (1.14a)$$

showing that the group of $\delta n(E)$ carriers "decays" or "relaxes" back to its thermal equilibrium state ($\mathbf{v}_i = 0$) with a time constant $\tau_c(E)$. What is occurring here is that *all* the members of the group $\delta n(E)$ with kinetic energy near E had an additional small velocity $\mathbf{v}_i(0)$ at $t = 0$; but with the passage of time an increasing number of these have their energy changed by collisions to a value outside that of the group, and they are replaced in the group by new members that have kinetic energy near E, but no incremental velocity component at all. Thus the amount of *organized* velocity *of the group* decreases with time. In another way of saying it, the number of particles in the group stays the same, so the *current* carried by the group in the direction of $\mathbf{v}_i$ simply decays with time. It is replaced by random current. The same reasoning applies to the net momentum of the group as to the current.

Next, for our introductory discussion of the effect of applied forces, let us consider the simple case in which the relaxation time τ_c

is the *same* for all kinetic energy groups; that is, τ_c is *independent* of E. Then, if the force $\mathbf{F}$ on the negatively charged conduction electrons is produced by an electric field $\mathbf{E}$, we know $\mathbf{F} = -q\mathbf{E}$. In the steady state, $d\mathbf{v}_i/dt = 0$ and the solution to Eq. 1.13*b* is just a steady *drift velocity* $\mathbf{v}_d$, independent of E

$$\mathbf{v}_d = \frac{-q\tau_c}{m_e^*}\mathbf{E} = -\mu_e\mathbf{E} \tag{1.14b}$$

If the velocity is given in cm/sec, and field in volts/cm, the proportionality constant μ_e has dimensions of cm²/volt sec.† The quantity μ_e is called the *mobility* (for conduction electrons) and is by convention always defined to be a positive number.

$$\mu_e = \frac{q\tau_c}{m_e^*} \tag{1.15}$$

Clearly the mobility is the incremental average speed per particle, per unit electric field (P1.6).

In this simple case, the relaxation time for *all* the electrons is the same as that of any group, namely, τ_c. Sometimes the time τ_c *in the mobility expression* Eq. 1.15 is called the *mean free time*. Such an idea comes from thinking about the collisions as if they were sudden discrete events, separated by absolutely "free times" between them. If the collisions occur properly at random in this kind of a model, we find a result like Eq. 1.15, in which the time that appears is just the average time between the idealized collisions. Inasmuch as we have seen that collisions do not really occur this way, all we shall use from such models is the name: mean free time.

Collisions of electrons with lattice atoms which are out of their equilibrium positions, because of thermal vibration, would be expected to happen more often at higher temperatures as the atoms develop larger amplitudes of vibration. Thus the mobility would be expected to *decrease* with increasing temperature when thermal vibrations dominate the scattering process. The situation for scattering by charged impurities is different. The deflecting effective-

† These units are the ones commonly used. But Eq. 1.14*b* and those based upon it (like Eq. 1.15) are treacherous because they mix electrical and mechanical variables. Thus they *must* be used in a *consistent set* of units. As mentioned before, in this book we have in mind mks rationalized units. See P1.10 and P1.15.

ness of the charged ion is greater, the lower the speed of the approaching charged carrier. Thus the scattering of electrons or holes by such ionized impurities is more effective at low temperatures. This means that mobility tends to have a broad maximum at intermediate temperatures. On the other hand, at low *enough* temperatures the donors and acceptors may not be ionized, so their scattering effect may eventually drop again. Some data confirming the aforementioned trends of the dependence of mobility upon temperature appear in Table 1.2 and Fig. 1.7 (P1.8 and Laboratory A2.3). The effect of charged impurities on mobility at normal temperatures is shown more clearly, for both majority and minority carriers, in Fig. 1.8 (P1.8, 1.9). At room temperature and above,

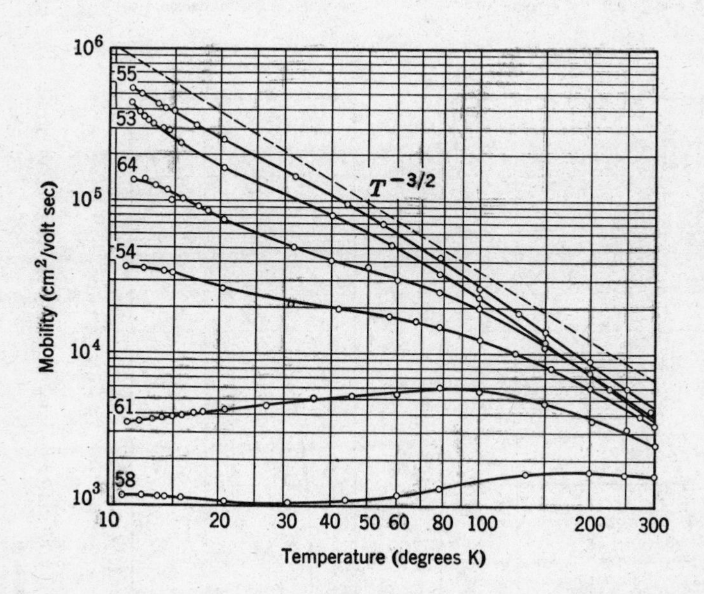

Fig. 1.7. Mobility versus temperature for a set of arsenic-doped germanium crystals. The numbers on the curves identify the samples as follows: No. 55, $N \equiv N_d - N_a = 1.0 \times 10^{13}$ cm^{-3} ($N_a < 2 \times 10^{12}$ cm^{-3}); No. 53, $N = 9.4 \times 10^{13}$ cm^{-3} ($N_a < 2 \times 10^{12}$ cm^{-3}); No. 64, $N = 1.7 \times 10^{15}$ cm^{-3} ($N_a < 2 \times 10^{13}$ cm^{-3}); No. 54, $N = 7.5 \times 10^{15}$ cm^{-3} ($N_a < 1.5 \times 10^{14}$ cm^{-3}); No. 61, $N = 5.5 \times 10^{16}$ cm^{-3} ($N_a < 5 \times 10^{15}$ cm^{-3}); No. 58, $N \sim 10^{18}$ cm^{-3}. The dashed line represents theoretical slope for scattering by lattice thermal vibrations (see P1.8). These data, and those of Figs. 1.11 and 1.13, were reported by E. M. Conwell, *Proc. IRE*, **40**, (Nov. 1952), pp. 1327–1337.

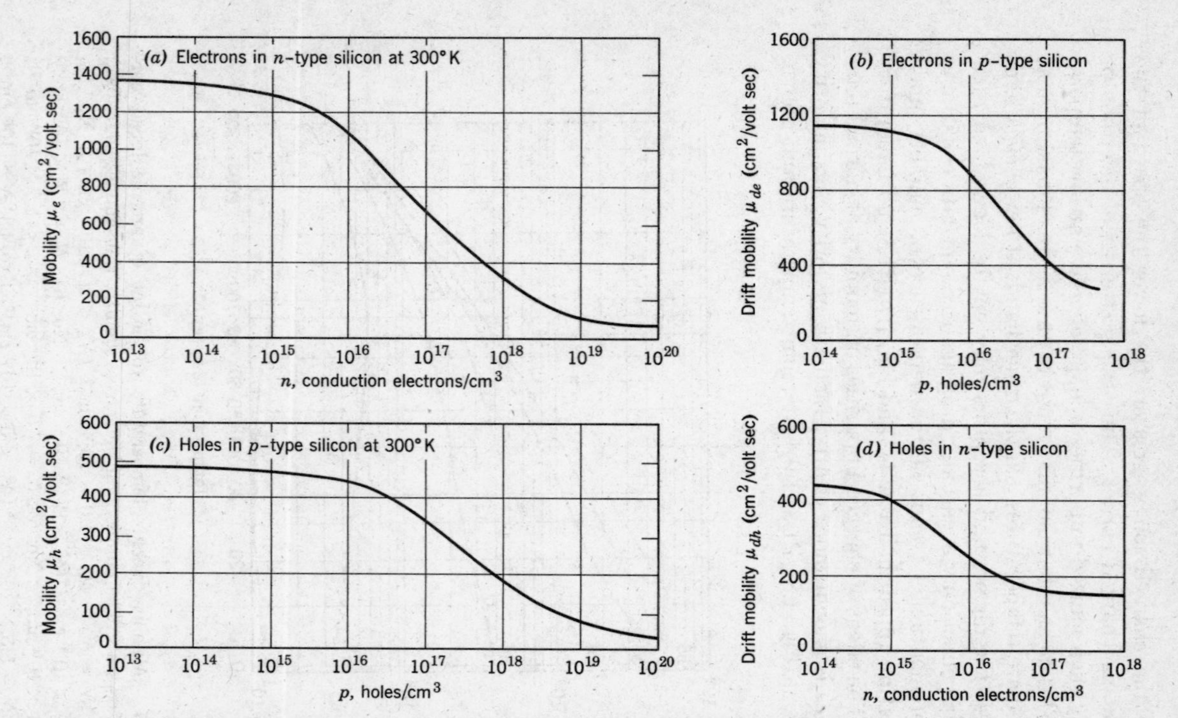

Fig. 1.8. Mobility of majority- and minority-carriers in silicon at room temperature, as functions of the majority-carrier concentration.

collisions with the vibrating atoms usually dominate, although the data shows that the impurities do interfere more as their concentration is increased. We shall soon understand better why the majority-carrier concentration (n and p for n- and p-type material, respectively) is a relatively direct measure of the impurity concentration, except at low values of n and p.

TABLE 1.2

Mobility of carriers in very pure germanium and silicon.†
(*Thermal vibration of the lattice should be by far the dominant scattering process*).

Material	Mobility at 300°K (cm²/volt-sec)	Approximate Temperature Dependence of Mobility
Ge		
Free electrons	3900	$4.9 \times 10^7 T^{-1.66}$ (100–300°K)
Holes	1900	$1.05 \times 10^9 T^{-2.33}$ (125–300°K)
Si		
Free electrons	1350	$2.1 \times 10^9 T^{-2.5}$ (160–400°K)
Holes	480	$2.3 \times 10^9 T^{-2.7}$ (150–400°K)

† E. M. Conwell, *Proc. IRE*, **46** (June 1958), pp. 1281–1300.

1.5.2 *A Relation between the Equilibrium Carrier Concentrations*

Besides the mobility, the concentration of carriers enters the problem of determining the electrical conductivity. This concentration is not a static concept, but may be looked upon as the average result of the balance between competing dynamic processes.

Imagine first a perfectly pure, intrinsic semiconductor, without any crystalline faults. Conduction electrons and holes are produced in pairs at a rate (number per unit volume, per unit time) that depends in part on the semiconductor material under consideration (i.e., the energy and momentum needed to produce a pair), and in part on the thermal activity of the surroundings. This activity shows itself as thermal vibration of the crystal lattice, and in the form of a general level of thermal radiation energy, both characteristic of the temperature T at which the material is in thermal equilibrium. To produce a hole-electron pair, some of the thermal energy must be channeled into breaking a valence bond.

For a given type of bond, with other ambient conditions held fixed, the chance of this happening per unit time depends only on the random features that go with the temperature. Let the rate at which bonds are broken in this *ideal* material be G_{ith} pairs/cm³-sec. The important point then is that, for a given *perfect*, intrinsic semiconductor material, $G_{ith}(T)$ depends *only* on temperature (other ambient conditions remaining fixed).

But matters become less obvious if we now ask what happens to $G_{ith}(T)$ when we add some donors or acceptors to the ideal material. Suppose we add *just donors*, at concentration N_d per cubic centimeter, and suppose these donors are all ionized at the temperatures of interest. In the usual case, N_d will be *much* smaller than the number of normal crystal atoms per unit volume, so it should not reduce appreciably the concentration of normal valence bonds available to be broken thermally, nor should it alter very much the thermal vibrations of the crystal lattice. If N_d is not too large, therefore, we would expect the rate at which bonds are broken thermally to be the *same* as it was in the intrinsic material at the same temperature; namely, $G_{ith}(T)$.

But opposing the process of pair *generation* described above is the simultaneous process of pair *recombination*. This action takes place when a conduction electron somehow loses its ionization energy and reenters a broken bond. In our idealized material, the number R_i of such recombination events per unit volume, per unit time (the recombination rate) will in general depend on the equilibrium conduction-electron concentration n_o, the equilibrium hole concentration p_o, the kind of basic semiconductor involved, and the temperature T. The equilibrium carrier concentrations n_o and p_o enter because, to recombine, a conduction electron must first find a hole, and this process occurs with a frequency that is dependent on the number of each present. Having encountered one another, however, the energy and momentum that the pair must give up in the recombination process depend on the bond structure of the particular material involved. The dynamical circumstances of the encounter, which govern how long it takes for the conduction electron to settle back into the broken bond by emitting radiation or transferring energy to the lattice vibrations, will on the average be set by the thermal motions of the carriers, and therefore by the temperature.

The presence of fully ionized impurities like donors, in very small concentrations, has practically no *direct* effect on recombination. Aside from preferentially increasing the concentrations of one or the other type of carrier, which is taken care of by the dependence of the chance of encounter on n_o and p_o, the very few impurity atoms involved do not significantly change the crystal environment in the neighborhood of enough of the normal bonds to affect the recombination dynamics in any appreciable number of recombination events. Therefore, for a group of samples of a given basic *ideal* semiconductor material, each of which may however contain different amounts of donors and/or acceptors over a wide range of concentrations (but all very small in atomic per cent), the recombination rate is the *same* function $R_i(n_o, p_o, T)$, in which the function R_i is characteristic of the *ideal intrinsic* material.

Now, in the state of thermal equilibrium, the concentrations of the carriers in any sample are not changing with time on the average. So, on the average, the pair generation rate must balance the pair recombination rate everywhere.

$$G_{i\text{th}}(T) = R_i(n_o, p_o, T) \tag{1.16a}$$

But, regarded as a function of n_o and p_o, at fixed T (i.e., for different samples of the same basic ideal semiconductor, characterized by different small amounts of donor and/or acceptor impurities, but all at the same temperature), we can make a Taylor expansion of R_i in two variables:

$$R_i(n_o, p_o, T) = a_i(T) + b_i(T)n_o + c_i(T)p_o + r_i(T)n_op_o + e_i(T)n_o^2p_o + f_i(T)n_op_o^2 + \ldots \tag{1.16b}$$

Here, however, it is clear that $a_i(T) = b_i(T) = c_i(T) \equiv 0$, because there can be no *pair*-recombination rate in a sample in which only *one* carrier type is present, and certainly none in a sample in which *neither* is present. So for the small carrier concentrations which are of greatest interest in electronic devices, the leading nonzero term of Eq. 1.16*b* is dominant and Eq. 1.16*a* becomes:

$$G_{i\text{th}}(T) = r_i(T)p_on_o \tag{1.17a}$$

or

$$p_on_o = \frac{G_{i\text{th}}(T)}{r_i(T)} \tag{1.17b}$$

Of course, Eq. 1.17*b* holds for intrinsic material as a special case, and there $p_o = n_o$. If we denote by n_i the concentration of either conduction electrons or holes in intrinsic material, it therefore follows from Eq. 1.17*b* that

$$\frac{G_{\text{ith}}(T)}{r_i(T)} = n_i^2 \tag{1.17c}$$

from which we draw the relatively obvious conclusion that, for a given ideal semiconductor, n_i is a function of T only, $n_i(T)$, and that in intrinsic material both carrier concentrations are identical functions of temperature $p_o = n_o = n_i(T)$.

Accordingly, Eq. 1.17*c* permits us to rewrite Eq. 1.17*b* in the form

$$p_o n_o = n_i^2(T) \tag{1.18}$$

The result is that an effort to "sprinkle in" conduction electrons by adding donors to a semiconductor, thus making $n_o > n_i$, has to result in *fewer* holes, $p_o < n_i$, in order to prevent the additional conduction electrons from recombining with holes faster than the unchanged thermal generation rate of pairs will permit.

It is clear that addition of acceptors alone can be discussed in the same manner and will again lead to Eq. 1.18.

The discussion so far has been phrased deliberately to omit consideration of recombination through imperfections in the lattice or at surfaces, which processes often in fact turn out to dominate the whole recombination phenomenon! In thermal equilibrium, however (to which, it must be emphasized, the above arguments are entirely restricted), it turns out that this apparent limitation of Eq. 1.18 to ideal material is entirely illusory. Here we are facing an important principle that underlies the very meaning of thermal equilibrium, and it is now useful to take it up explicitly.

1.5.3 *Principle of Detailed Balance*

In thermal equilibrium, every identifiable physical process that takes place proceeds *on the average* at exactly the same rate as its own inverse. One physical process cannot be compensated for by a different one; each must *self-balance* independently of the others.

Known as the *principle of detailed balance*, this idea depends on the second law of thermodynamics in the following manner. Suppose we consider, as an example, two different ways of producing and recombining hole-electron pairs. Let one be a process that either gives off or absorbs a photon of light to account for the ionization energy. Let the other be a different process that either gives up to or takes on mechanical energy from the crystal lattice vibrations, without involving any light at all. Assume that in unit time, under equilibrium conditions, the first process produces more recombinations than generations of carrier pairs, whereas the second one just compensates the energy by producing more generations than recombinations. If these processes have real physical meaning, we can observe them, and in that event we can collect the excess light emitted by the first process and prevent it somehow from being returned to the lattice. Then that process will just be balanced, but the second one will continue to produce more pairs than it recombines. The net mechanical energy it thus takes out of the lattice vibrations must result in a *cooling* of the crystal. In this way, from a crystal lying on the table at some equilibrium temperature T, we can extract photons to do some useful work (like operating a photovoltaic light meter, for example), and the only price we pay is that the sample on the table cools itself off spontaneously! Or, if you wish, we can use our photons to heat up a different sample lying on the same table, and have a situation in which one sample spontaneously gets cooler while the other one gets hotter, and we have done no work at all! It is just this kind of situation which is prohibited by the second law of thermodynamics, and any physically meaningful violation of the principle of detailed balance may be reduced to such a violation of this law.

Hence, our entire argument in Sec. 1.5.2 may be recast verbatim for the general case, simply by *redefining* R_i and G_{ith} to be, respectively, the recombination and generation rates for *only* that subclass of recombination-generation processes that do *not* involve *any* imperfections. Thus the result 1.18 is correct as an equilibrium property of the whole system, and could in principle have been derived from a detailed consideration of the balance of *any* of its sub-processes. Indeed Eq. 1.18 is also true even if the impurity atoms are *not* all ionized, inasmuch as the sub-processes considered in deriving it are unaffected by the state of ionization of the impurities.

Results like Eq. 1.18 often occur in chemistry, where they are known as laws of *mass action* and have to do with the extent to which a reaction rests one way or the other at equilibrium. We can phrase our present problem in the form of a pseudo-chemical process:

$$\text{Conduction electron} + \text{hole} \underset{+E_g}{\overset{-E_g}{\rightleftharpoons}} \text{unbroken bond}$$

indicating that the ionization energy E_g has to be disposed of. We then notice that the term $n_i^2(T)$ in Eq. 1.18 would on this basis involve something like the very large concentration of unbroken bonds, and a factor $e^{-E_g/kT}$. In Chapter 3 we shall find these same features in the present problem.

1.5.4 *Equilibrium Carrier Concentrations in Homogeneous Semiconductors*

Given the intrinsic concentration n_i of conduction electrons or holes in a parent semiconductor, like Ge, at a temperature high enough so that any excess donors or acceptors in the material will be completely ionized, Eq. 1.18 furnishes *one* of the relations needed to determine the new hole and conduction-electron concentrations in the presence of "doping" impurities.

For the special case of *homogeneous* material containing *both* donors and acceptors, an additional relation comes from the fact that in every volume of such material large enough to contain many atoms of all the types involved, every conduction electron must have come from either a broken bond or an ionized donor, and every hole must be accounted for by either an ionized acceptor or the broken bond left by one of the conduction electrons. Thus, if we subtract the concentration of ionized donors (assumed to be *all* of them, N_d, for our present purposes) from the concentration of conduction electrons n_o, and also subtract from p_o the ionized acceptor concentration N_a, we should be left with only hole-electron pairs. The resulting numbers of conduction electrons and holes should be equal:

$$\left.\begin{aligned} &(a)\quad n_o - N_d = p_o - N_a \\ \text{or}\quad &(b)\quad N_a - N_d = p_o - n_o \\ \text{or}\quad &(c)\quad n_o + N_a = p_o + N_d \end{aligned}\right\} \qquad (1.19)$$

If multiplied by the electronic charge q, the second form of Eq. 1.19 states that the *net positive mobile charge* $q(p_o - n_o)$ is equal in magnitude to the *net negative immobile charge* $q(N_a - N_d)$, if we recall that ionized acceptors are negatively charged and ionized donors are positively charged. Similarly, the third form of the equation *equates the total fixed and mobile negative charge to the corresponding total positive charge*, for each unit volume of the uniform material. In terms of what we shall refer to as the "excess" donor concentration $N = N_d - N_a$, which may be positive or negative, Eq. 1.19 reads

$$n_o - p_o = N \tag{1.20}$$

An alternate physical interpretation of the result in Eq. 1.20 proceeds as follows. If both donors and acceptors are present simultaneously in a semiconductor at normal temperatures, they will all be ionized just as if they were present alone. An excess electron on a donor is still ready to become completely free with very little addition of energy, and an electron filling the excess bond on an acceptor is still only slightly less strongly held than one in an ordinary valence bond. This situation suggests that the donors supply N_d conduction electrons once and for all, and the acceptors supply N_a holes once and for all. But, clearly, if N_d and N_a are unequal, the free carriers supplied by the least numerous will simply recombine *once and for all* with an equal number of the carriers supplied by the most numerous. This leaves only those carriers supplied in effect by the excess $N = N_d - N_a$ as those "sprinkled in" by the impurities to affect the equilibrium balance between pair generation and pair recombination. If N is positive, it represents the *effective* concentration of donors sprinkling in conduction electrons. If N is negative, acceptors are more numerous than donors, and N represents in magnitude the *effective* acceptor concentration capable of furnishing holes at ordinary temperatures.

Provided all the impurities are ionized, Eqs. 1.18 and 1.20 are sufficient to determine p_o and n_o, if $n_i(T)$ is known for the pure parent semiconductor. This function happens to be exponential in form, and numerical data pertaining to it appear in Eqs. 3.59 and 3.60, for germanium and silicon. The same data are presented graphically for easier use in Figs. 1.9 and 1.10.

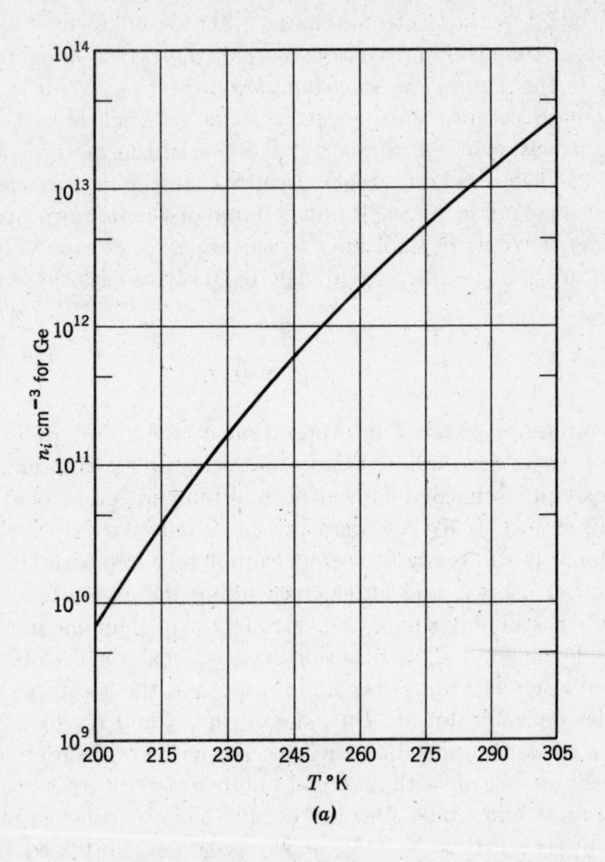

Fig. 1.9. Intrinsic carrier concentration as a function of temperature for germanium (Eq. 3.59*a*): (*a*) 200° to 305°K.

It is clear from Eq. 1.20 that if $N \gg n_i$, $n_o \gg n_i$. Accordingly, from Eq. 1.18 we see that $p_o \ll n_i$. It follows from Eq. 1.20 again that p_o is negligible there, and

$$n_o \approx N.$$

As long as the excess impurities are ionized and N remains large compared to n_i, the conduction-electron concentration n_o remains approximately constant with temperature. The hole concentration, however, is given approximately by Eq. 1.18 as

$$p_o = \left(\frac{n_i}{n_o}\right) n_i \approx \left(\frac{n_i}{N}\right) n_i \ll n_i \quad \text{if} \quad N \gg n_i$$

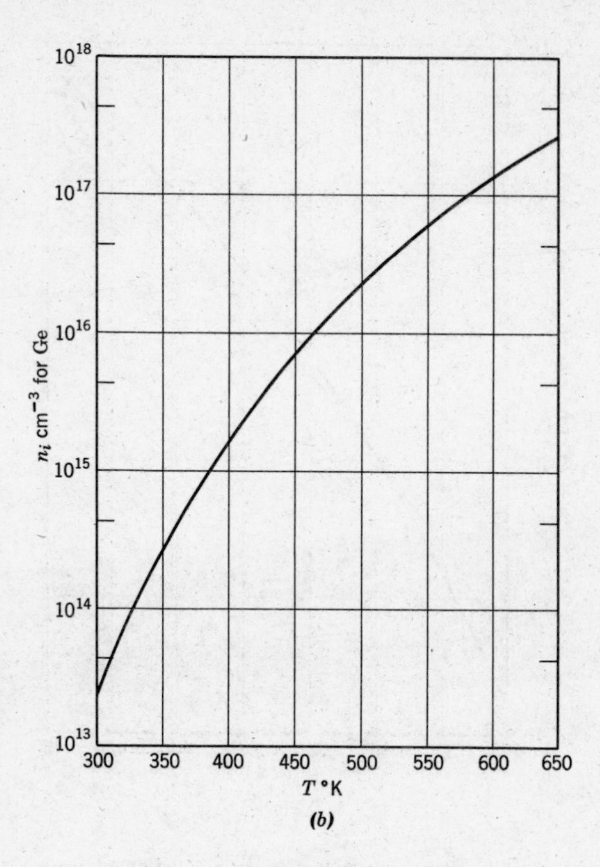

Fig. 1.9. (*Continued*) (*b*) 300° to 600°K.

which is small, but is the same violent (exponential) function of temperature as $n_i^2(T)$ (see Eq. 3.57). The condition $N \gg n_i(T)$ may occur at low enough temperatures in a given n-type material, or for a large enough donor concentration at a given temperature.

Similarly, in p-type material, for which $N < 0$, the temperature and acceptor concentration may be such that $|N| \gg n_i$. Then from Eq. 1.20 we find $p_o \gg n_i$, and from Eq. 1.18 $n_o \ll n_i$. It then follows from Eq. 1.20 again that n_o is negligible, and

$$p_o \approx |N|$$

virtually independent of temperature, and from Eq. 1.18 that

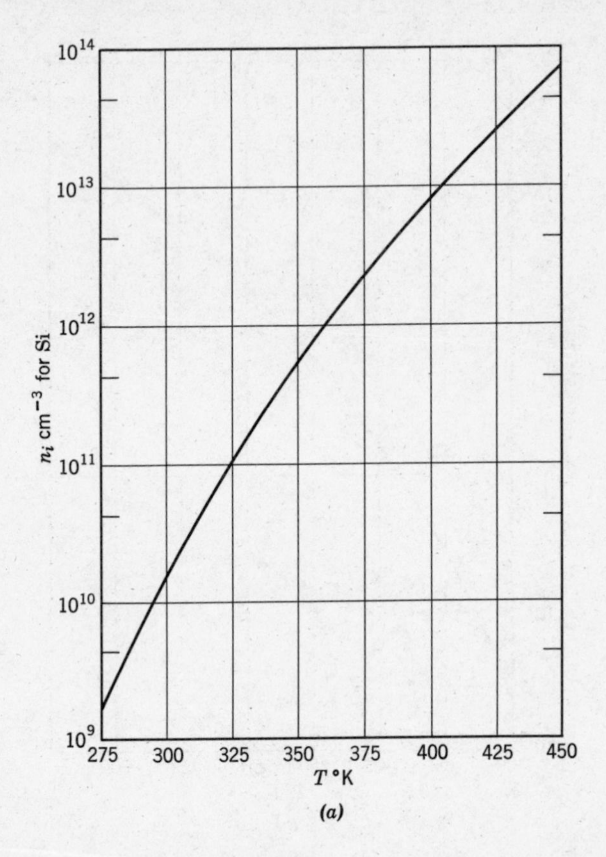

Fig. 1.10. Intrinsic carrier concentration as a function of temperature for silicon (Eq. 3.59*b*): (*a*) 275° to 450°K.

$$n_o \approx \frac{n_i^2}{N} = \left(\frac{n_i}{N}\right) n_i \ll n_i$$

which varies with temperature like $n_i^2(T)$.

Under the two circumstances described above, in which $|N| \gg n_i$, the semiconductor in question is said to be *extrinsic* n-type or p-type (as the case may be). Its *majority*-carrier concentration is large and nearly equal to the excess impurity concentration $|N|$, virtually independent of temperature. The *minority*-carrier concentration is small and varies [like $n_i^2(T)$] almost exponentially with temperature.

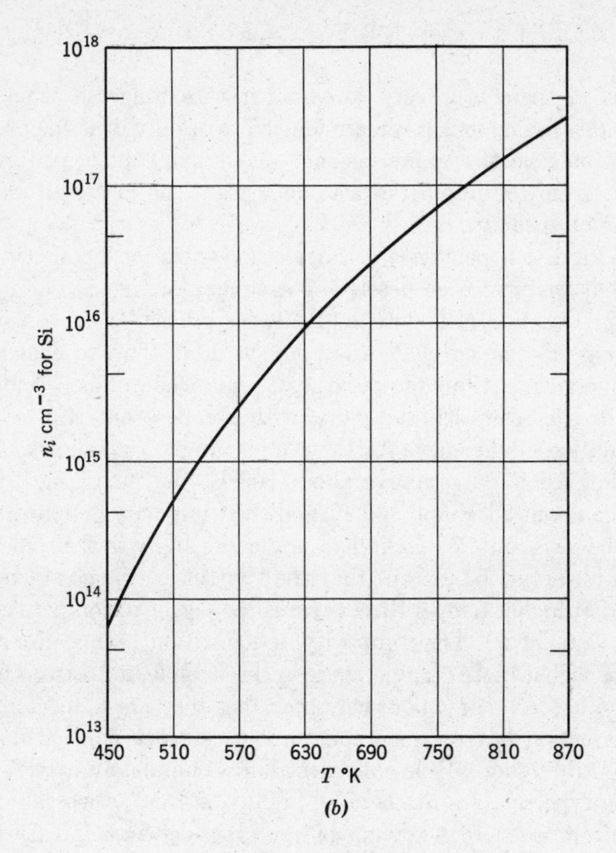

Fig. 1.10 (*Continued*) (*b*) 450° to 870°K.

If we consider a given sample at various temperatures, the temperature range in which all the impurities are ionized and $|N| \gg n_i$ is called the *extrinsic temperature range* (sometimes also called the "saturation" or "exhaustion" range).

Eventually, however, at high temperatures n_i is bound to become comparable to $|N|$. The number of hole-electron pairs generated from broken bonds competes with the carriers supplied by the impurities. Both p_o and n_o become large compared to $|N|$ when $n_i(T) \gg |N|$, and therefore $p_o \approx n_o \approx n_i(T)$. We often say that the semiconductor has entered the *intrinsic temperature range* under these conditions, although it is certainly no "purer" than it was before.

Finally, there is a very low-temperature range in which the impurities can no longer remain ionized. Not only does the production of hole-electron pairs become very small, but the supply of carriers from the impurities also decreases. The material tends to become an insulator.

The kind of dependence of carrier concentration upon temperature that we have been describing is shown for the case of germanium in Fig. 1.11 (P1.11 to 1.14). These particular curves, which show log (carrier concentration) versus $1/T$, tend to exaggerate the phenomena at low temperatures, where the number of ionized impurities is decreasing rapidly with temperature. But careful examination of the curve for the purest sample (55) shows clearly the "intrinsic" range above about 300°K, the "extrinsic" range between about 25°K and 300°K, and the "impurity de-ionization" range below about 25°K. Similar comments apply to the somewhat less pure sample 53, except that the "intrinsic" range starts at a somewhat higher temperature (n_i must be bigger to compete with a larger value of N). The "impurity de-ionization" range also starts at a somewhat higher temperature in this sample, for reasons which we need not say more about here than that they are connected with a mass action law between conduction electrons, ionized donors, and neutral donors. Hole concentration is completely negligible in these n-type samples at these low temperatures, as we can easily prove from Eq. 1.18, the value of n_o on these curves, and the value of $n_i^2(T)$ from Eq. 3.57a. The same trends continue through the successively more heavily doped samples 64, 54, and 61. [The unexpected small dip near the intrinsic line, especially in 64, is not actually a property of the carrier concentration, but rather the result of unavoidable complications in the Hall effect involved in these measurements, at temperatures near the intrinsic range (P1.19 and P1.20).] The very impure sample 58 departs radically from the theory given here; the nature of this problem is discussed briefly in Sec. 2.3.

1.5.5 *Conductivity*

To understand the conduction process, consider a portion of a one-carrier conductor (Fig. 1.12). First, let there be *no* applied

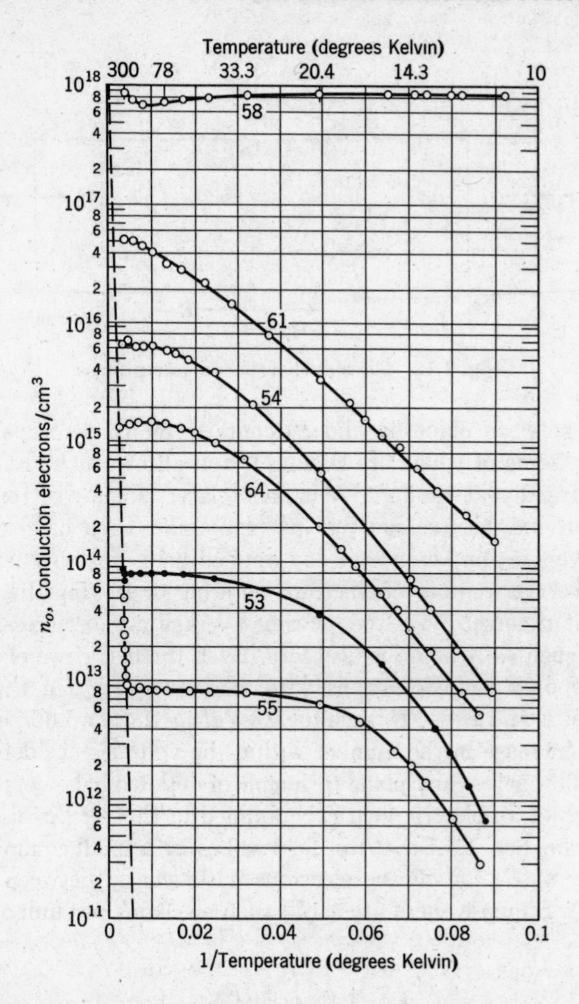

Fig. 1.11. Concentration of charge carriers versus reciprocal absolute temperature for the same arsenic-doped germanium samples shown in Fig. 1.7. The dashed line represents the concentration of intrinsic carriers $n_i(T)$.

field and assume that the conductor is in thermal equilibrium. Under these circumstances let there be n conduction electrons per m³ in the conductor, on the average. This number is fixed by the atomic and crystal structure, along with the neutrality requirement. A given macroscopic volume of the conductor (e.g., that

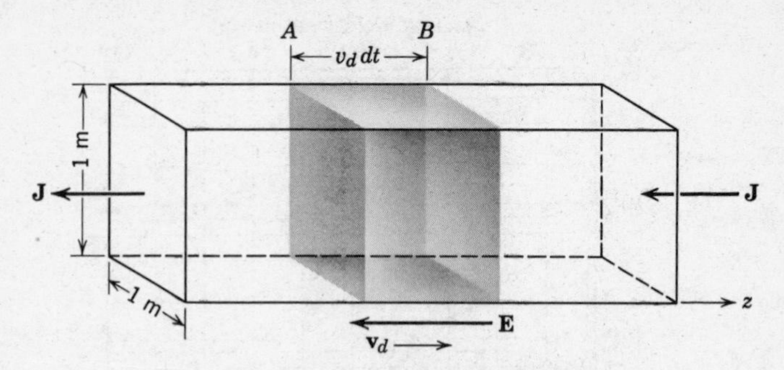

Fig. 1.12. The electronic conduction process.

shown between planes A and B) contains many electrons at any instant, some of which are leaving it and all of which are moving about rapidly at random. A moment later, some electrons have left; but on the average just as many new ones have entered. Moreover, in the absence of any applied field, not only is the *net* time-average number of electrons entering or leaving the volume per unit time zero, but the *net* time-average number crossing any plane (such as A or B) is *also* zero. Even further, if we elected to observe only those electrons with thermal energy in the range between E and $E + dE$, then for *these alone* there would be no net rate of increase in the number within the volume considered, and no net flow across any plane (principle of detailed balance).

Next, let an electric field $\mathbf{E}$ be applied in the $-z$ direction. We know from Sec. 1.5.1 that this field will cause a small common drift velocity $\mathbf{v}_d(E)$ for *all* the electrons with energy between E and $E + dE$. Suppose there are $\delta n(E)$ such electrons per unit volume, so that

$$n = \int_{\text{all } E} \delta n(E) \tag{1.21}$$

Then if we again consider a macroscopic volume of the conductor, like that between planes A and B of Fig. 1.12, the net number of electrons in it which have energies near E will still not change with time, on the average. There *is* now, however, an average drift of such electrons *into* the volume across plane A, and equally *out* of it across plane B, because of the organized drift velocity $\mathbf{v}_d$.

To determine the amount of charge crossing a plane per unit time, let the distance between planes A and B be $dz = v_d\,dt$. Then the average number of electrons which have energy near E and are between the planes is just $\delta n(E)v_d\,dt$. Because all of these have the same drift component $\mathbf{v}_d$, all of them on the average cross plane B from left to right in time dt, all other random motions across plane B being cancelled out on the average by the thermal velocities of particles on both sides of the plane.

Noting that each electron carries a charge $-q$, and that the cross-sectional area in Fig. 1.12 is 1 m^2, we recognize that the charge per unit time crossing plane B has the dimensions of amperes per square meter. It is referred to as a *current density*, with symbol $\mathbf{J}$. (Actually, in the general case, plane B is defined as that one which is oriented to intercept the greatest flow; the current density $\mathbf{J}$ is then a vector normal to that plane, of magnitude equal to the flow rate.) Thus the group $\delta n(E)$ of conduction electrons considered here contributes to the current density an amount

$$\delta\mathbf{J} = \frac{-q\mathbf{v}_d(E)\delta n(E)\,dt}{dt} = -q\mathbf{v}_d(E)\,\delta n(E) \tag{1.22}$$

The total current density produced by all energy groups is obtained by adding (integrating) all the separate contributions

$$\mathbf{J} = -\int_{\text{all } E} q\mathbf{v}_d(E)\delta n(E) \tag{1.23}$$

in which we know from Eq. 1.13b that in the steady state $[(d\mathbf{v}_i/dt) = 0]$ there is a steady drift velocity $\mathbf{v}_d(E)$:

$$\mathbf{v}_d(E) = -\frac{q\tau_c(E)}{m_e^*}\mathbf{E} \tag{1.24}$$

So

$$\mathbf{J} = \frac{q^2}{m_e^*}\left[\int_{\text{all } E} \tau_c(E)\,\delta n(E)\right]\mathbf{E} \tag{1.25}$$

Let us now define and use the average relaxation time $<\tau_c>$ for all the conduction electrons, of which we have said there are n per unit volume (Eq. 1.21):

$$<\tau_c> \equiv \frac{\displaystyle\int_{\text{all } E} \tau_c(E)\delta n(E)}{\displaystyle\int_{\text{all } E} \delta n(E)} = \frac{1}{n}\int_{\text{all } E} \tau_c(E)\delta n(E) \tag{1.26}$$

On this basis Eq. 1.25 becomes

$$\mathbf{J} = q\left(\frac{q<\tau_c>}{m_e^*}\right)n\mathbf{E} \equiv q\mu_e n\mathbf{E} \tag{1.27}$$

where now we have put

$$\mu_e \equiv \frac{q<\tau_c>}{m_e^*} \tag{1.28}$$

In other words, the mobility is related to the *mean* relaxation time in such cases (compare Eq. 1.15). One should be cautioned, however, that the result of solving Eq. 1.13*b* for each energy group, and then adding up the results, or averaging, does *not* always simply replace τ_c by $<\tau_c>$.

Notice that Eq. 1.27 may be written

$$\mathbf{J} = (q\mu_e n)\mathbf{E} = \sigma_e\mathbf{E} \tag{1.29}$$

which is precisely Ohm's law, with σ_e the conductivity of the material (mhos per m) and E the voltage drop per meter! (P1.15).

In a semiconductor there are two kinds of mobile carriers present at the same time. Because they are oppositely charged, they move in opposite directions in an electric field. But, again because they are oppositely charged, they carry current in the *same* direction. The conduction electrons and holes have different effective masses, and they may be scattered differently. Therefore, they would not be expected to have the same mobility. Using subscripts e and h for conduction electrons and holes, respectively, the addition of currents for the two carriers yields:

$$\mathbf{J} = q(\mu_e n + \mu_h p)\mathbf{E} = (\sigma_e + \sigma_h)\mathbf{E} = \sigma\mathbf{E} \tag{1.30}$$

where evidently

$$\sigma = \sigma_e + \sigma_h = q(\mu_e n + \mu_h p) \tag{1.31}$$

Figure 1.13 shows the temperature dependence of the conductivity for the same samples involved in Figs. 1.11 and 1.7 (P1.16, P1.17). Observe that in the intrinsic range the conductivity follows essentially $n_i(T)$ which varies so very fast. Note also that in the extrinsic range the conductivity *rises* as temperature *drops*, even though the carrier concentration tends to remain constant (Fig. 1.11). This behavior reflects the increase in mobility (Fig. 1.7) with decreased scattering of carriers by thermal vibration of the crystal lattice. Eventually, however, the number of carriers drops

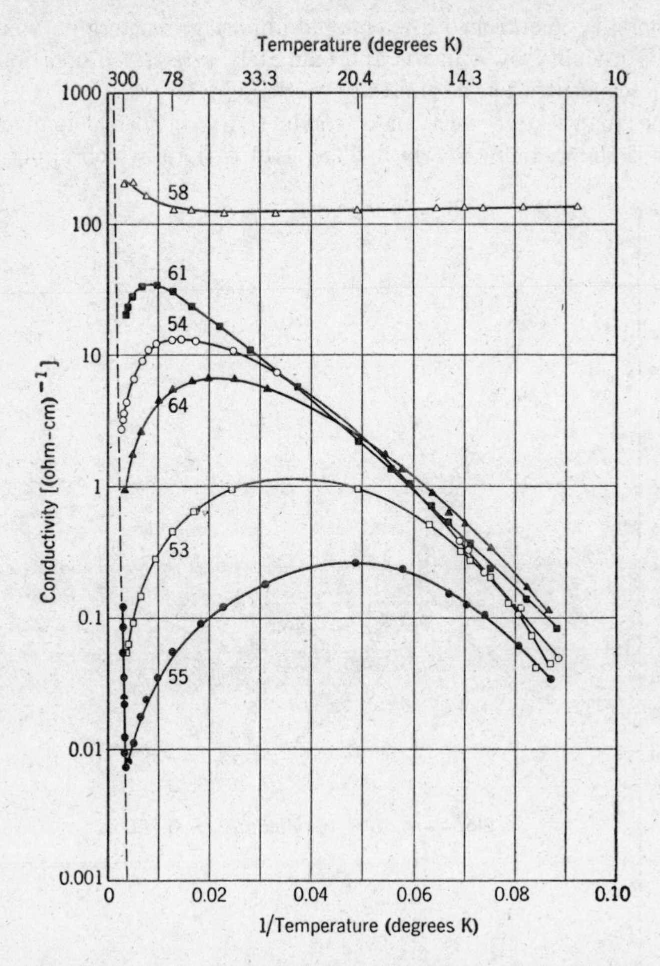

Fig. 1.13. Conductivity versus reciprocal absolute temperature for the same arsenic-doped germanium samples shown in Figs. 1.11 and 1.7. The dashed line represents the intrinsic conductivity.

as the impurities de-ionize, and also the mobility begins either to stop increasing so fast, or actually to decrease (Fig. 1.7), as a result of the scattering of carriers by charged impurities. Notice in fact that samples 53, 64, 54, and 61 cluster in conductivity at low temperatures, which indicates that, whereas those with high impurity content have larger values of carrier concentration (equal to the concentration of *ionized* impurities), they also have *smaller* values of mobility (controlled by charged-impurity scattering, which affects mobility by a factor approximately *inversely* proportional to the concentration of ionized impurities).

The quantitative behavior of conductivity at normal temperatures is shown more clearly in Figs. 1.14 to 1.16, which apply to

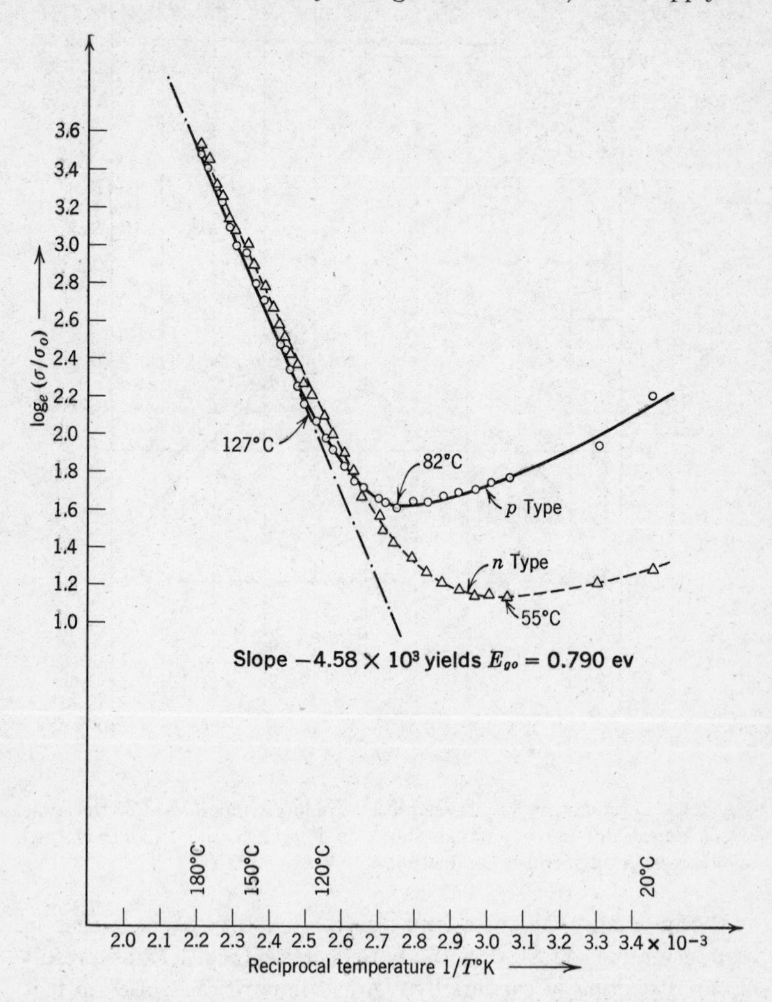

Fig. 1.14. Log of normalized conductivity versus reciprocal temperature for two germanium samples in the moderate (extrinsic) and high-temperature (intrinsic) ranges. The resistivity of both samples at room temperature is about 3 ohm-cm. (See also Figs. 1.15 and 1.16.)

Ge samples used in Laboratory A2.3. Figure 1.14 shows the log of normalized conductivity versus $1/T$ in the intrinsic *and* extrinsic regions. Figure 1.15 shows the intrinsic region in more detail (plotted as ln (resistance) instead of ln (conductance) versus $1/T$). This plot does show the expected exponential behavior of n_i versus $1/T$ in Eq. 3.59*a*. On the other hand, Fig. 1.16 shows the

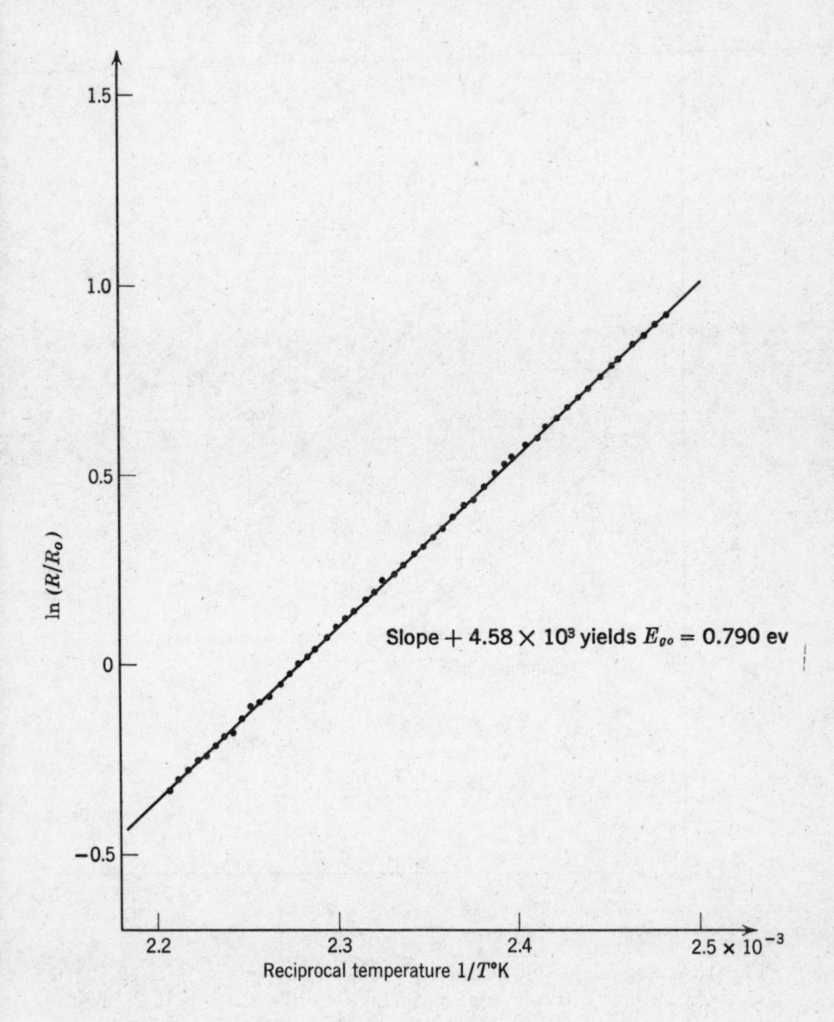

Fig. 1.15. Log of sample resistance versus $1/T$°K in the intrinsic range. Samples are the same as those used for Fig. 1.14

extrinsic range in detail plotted as ln (conductivity) versus ln(T). Here the power law expected for mobility variation with temperature should show up, as it indeed does (compare Table 1.2).

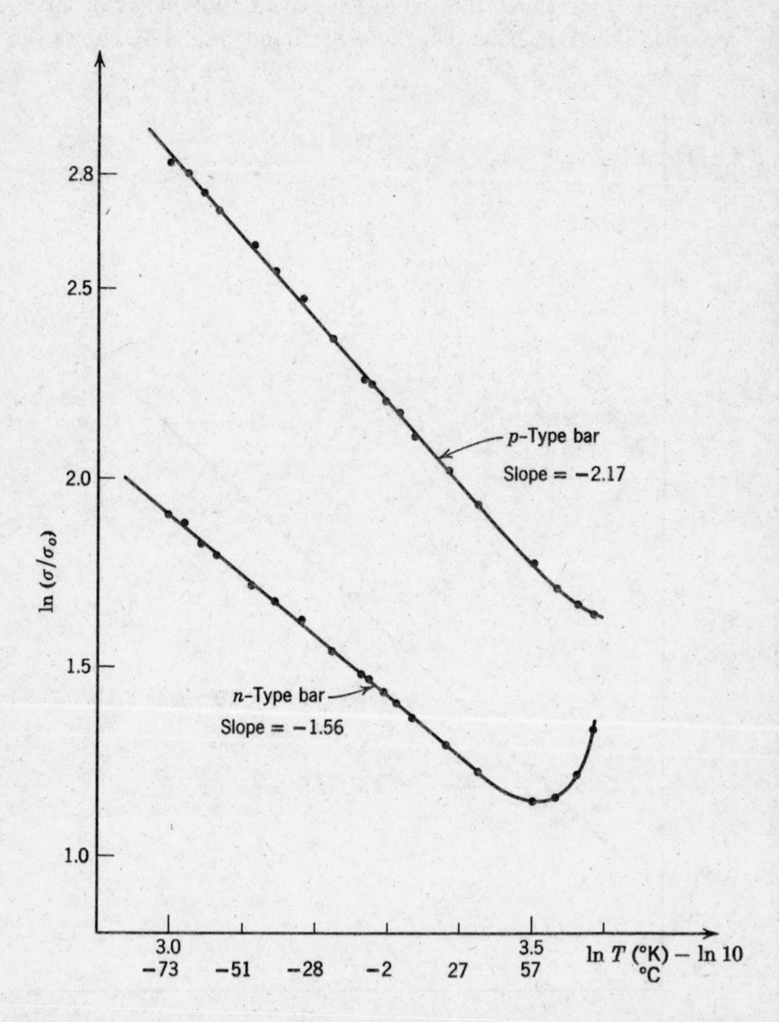

Fig. 1.16. Log-log plot of conductivity versus temperature, primarily for the low-temperature (extrinsic) region. Samples are the same as those used for Fig. 1.14.

1.5.6 *Hall Effect*

If a dc magnetic field B is imposed upon a current-carrying sample in the manner of Fig. 1.17, the carriers will be acted upon by both a steady electric field and a steady magnetic field. In the dc steady state, for the simple case of an extrinsic n-type material having a negligible hole population and a constant value of τ_c for all groups, Eq. 1.13*b* becomes

$$-q\mathbf{E} - q\mathbf{v}_d \times \mathbf{B} = \frac{m_e^* \mathbf{v}_d}{\tau_c} \tag{1.32}$$

or

$$\mathbf{E} = -\frac{m_e^*}{q\tau_c}\mathbf{v}_d - \mathbf{v}_d \times \mathbf{B} \tag{1.33}$$

But

$$\mathbf{J} = -nq\mathbf{v}_d$$

which can be used to eliminate $\mathbf{v}_d$ from Eq. 1.33, to yield

$$\begin{aligned}\mathbf{E} &= \left(\frac{1}{\mu_e nq}\right)\mathbf{J} + \left(\frac{1}{nq}\right)\mathbf{J} \times \mathbf{B} \qquad (1.34)\\ &= \rho_e\mathbf{J} - R_e\mathbf{J} \times \mathbf{B}\end{aligned}$$

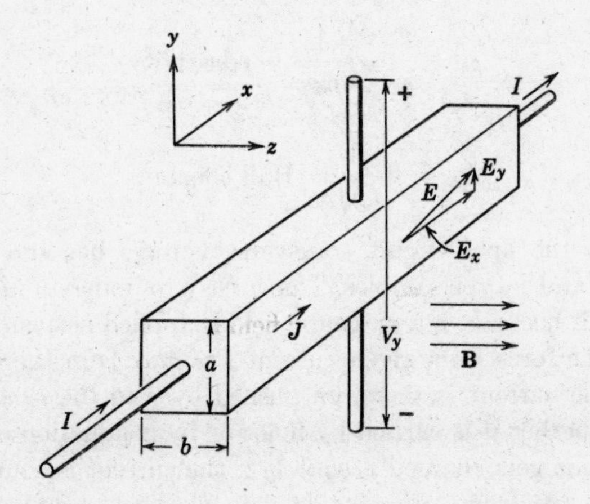

Fig. 1.17. Schematic arrangement for Hall effect.

where

$$\rho_e = \frac{1}{\sigma_e} = \frac{1}{q\mu_e n} = \text{resistivity}$$

and

$$R_e = -\frac{1}{nq} = \text{Hall constant} \tag{1.35}$$

On this simple picture, with only one carrier of importance and a single relaxation time τ_c for all thermal groups, we notice from Eq. 1.34 that the component of **E** along **J** is the same whether or not **B** is present. That is, the magnetic field does not affect *this component* of **E**. Also if **B** is parallel to **J**, it has no effect on the resistance of the sample at all (in this model, there is no "magnetoresistance" effect).

But if **B** is perpendicular to **J**, there is a component of **E** in the direction of $\mathbf{J} \times \mathbf{B}$, perpendicular to both **B** and **J**. That is, in Fig. 1.17 there is an open-circuit voltage developed *transverse* to the bar. This "Hall voltage" is linearly proportional to both **J** and **B** in this simple case, and the proportionality constant in Eq. 1.35 depends only on the carrier concentration n (and the electron charge q). In fact, if we carry out the same analysis for a p-type extrinsic bar we find that

$$\mathbf{E} = \rho_h \mathbf{J} - R_h \mathbf{J} \times \mathbf{B} \tag{1.36}$$

with

$$\rho_h = \frac{1}{\sigma_h} = \frac{1}{q\mu_h p} = \text{resistivity} \tag{1.37}$$

but now

$$R_h = +\frac{1}{pq} = \text{Hall constant} \tag{1.38}$$

Evidently, the open-circuit transverse voltage has the opposite sign for n- and p-type samples. This is easy to understand in terms of Fig. 1.17, because, if a magnetic field is applied perpendicular to the bar, the force on a given current **J** is *always* in the direction $\mathbf{J} \times \mathbf{B}$. The current is therefore pushed over to the *same* side of the bar, whether it is carried by holes or by conduction electrons. But that side gets charged *negatively* if the current is composed of conduction electrons, whereas it gets charged *positively* if the current is composed of holes.

By reason of its sign properties, and the direct connection of the Hall constant with the majority-carrier concentration in extrinsic material (Eqs. 1.35 and 1.38), the Hall effect is of great utility in the experimental characterization of semiconductors (Laboratory A2.4 and P1.6*c* and *d*, P1.19, and P1.20).

1.6 OTHER ELECTRONIC PROCESSES

1.6.1 *Minority-Carrier Injection*

We have spoken of the fact that atomic thermal vibrations can produce conduction electron-hole pairs in a semiconductor by giving enough mechanical energy to some of the bound electrons to set them "free" in the crystal. Essentially the same process may be produced by a light beam (photons). Of course, as in all photoelectric effects, the frequency of the light must be in the right range; i.e., the energy of its photons must be sufficient to break a covalent bond. The semiconductors of interest to us are very photosensitive to light in the near-infrared and the visible regions of the spectrum, a fact which forms the basis for their application in crystal photocells.

Suppose we have a homogeneous bar of semiconductor and irradiate it *uniformly* with light of sufficient frequency to produce conduction electron-hole pairs. Assume that the light also penetrates the sample uniformly, although the fact that it gets used up in producing carrier pairs puts limitations on the validity of this assumption. Imagine that the light is pulsed on very intensely, for a *very short* time. During the time the light is on, its intensity defines the number of photons per unit time entering the semiconductor and, accordingly, the *time-rate* at which carrier pairs are being produced by the light. Thus, just after the pulse, a certain number of hole-electron pairs have been produced and this number is in excess of that characteristic of thermal equilibrium at the temperature of the bar. We have a uniform electrically neutral cloud of mobile carriers, in which *both* carrier concentrations have been increased by equal amounts. If n and p are the new concentrations, and n_o and p_o are those characteristic of thermal equilibrium, the *excess* concentrations may be denoted by $p' \equiv p - p_o$ and $n' \equiv n - n_o$. In the present situation $n' = p'$. If the original bar was strongly n-type, so in equilibrium $n_o \gg p_o$ and $p_o \ll n_i$, the increase in hole concentration (minority carrier) produced by the light may be very great on a percentage basis, whereas the conduction-electron concentration (majority carrier) may increase by only a very small percentage—in spite of the numerical equality between the increments in concentration. It is customary to refer to this pair-creation process as *minority-carrier injection*, to focus attention on the dominant percentage effect.

It cannot be too strongly emphasized that almost all cases of "minority-carrier injection" are in fact pair-injection processes which create neutral clouds of both types of mobile carriers. We shall have many opportunities to deal with injection by means other than light, as this injection process is fundamental to the operation of almost all semiconductor devices, and it will pay us in every case to look for the electrical neutrality.

1.6.2 *Recombination*

Returning to our illuminated bar, we may inquire what happens to the excess carrier pairs after the light flash. The answer is obvious. They recombine in pairs until thermal equilibrium is established. The part of the question that is more difficult concerns the mechanisms of such recombination and the rates at which they proceed.

We can quite easily imagine two ways for a conduction electron and a hole to lose the energy (and possibly momentum) necessary to get a valence bond refilled; one way is by losing mechanical energy to the crystal lattice vibrations; another way is by giving off electromagnetic energy (light). It turns out, however, that the crystal lattice has so much inertia that it is hard to give it the entire ionization energy all at once, in a short time. If there happen to be impurities (other than donors or acceptors) in the crystal at which an electron can orbit for a moment with an energy intermediate between the "free" and "bound" conditions, a two-step process becomes possible, each step requiring the lattice to take less energy. Such impurities are in fact hard to avoid, as a practical matter of crystal growth, and simple faults in the regularity of the structure can readily furnish them or cluster them together. Surfaces are a very prolific source of such "recombination centers." This process of recombination, by using crystal faults to catalyze mechanical energy transfer, is usually the dominant one at normal temperatures and for the common device materials.

But the emission of electromagnetic energy is not to be passed over completely by any means. The main problem with this process is the *lack* of much inertia in light; so in those materials that require much momentum (as well as energy) change when an electron goes from "bound" to "free," a photon of light with the right energy just cannot handle the momentum. In these materials, mixtures of the mechanical and electromagnetic processes may occur in a recombination, and these combined processes can be significant.†

Some other semiconductor materials do not require much momentum for this purpose, however, and light radiation is then a major recombination process.‡

Finally, one other way for a pair to recombine is to transfer the energy and momentum to another carrier. Thus two carriers of one type and one of the opposite type can come together, with the third particle carrying away the recombination products of the other two. Such processes obviously demand ready availability of clusters of mobile carriers, and as such only become important when the "injection level" becomes very high.

When the *total concentrations* n and p of both types of carriers are *not too large*, we can expect the space-time rate of recombination of pairs, $R(n, p, T)$ pairs per unit volume per unit time, to obey the same kind of a general law discussed in connection with Eq. 1.17*a*, namely,

$$R(n, p, T) = r(T)np \tag{1.39}$$

This says simply that the recombination of a conduction electron with a hole is more probable in simple proportion to the number of

† An SEEC 16-mm, 40-minute, sound film relevant to this matter, entitled "Gap Energy and Recombination Light in Germanium," has been made by J. I. Pankove and R. B. Adler. Pending completion of commercial distribution arrangements, it may be borrowed for preview or purchased from the Film Librarian, Educational Services, Inc., 47 Galen Street, Watertown 72, Mass. See also Problems P2.4 and P3.2.

‡ In these cases, not only may there be a great deal of radiation, but it may even be made coherent. This phenomenon then becomes the basis for the injection laser.

holes, and that the number of such recombinations per unit time is proportional to the number of conduction electrons.

Now, however, we are considering nonequilibrium processes in which n and p may change in a *given sample* because of addition of n' and p' by injection,

$$n = n_o + n' \tag{1.40a}$$

$$p = p_o + p' \tag{1.40b}$$

and we are really interested in *all* the ways in which the *excess* carriers may recombine. Thus $r(T)$ in Eq. 1.39 differs from $r_i(T)$ in Eq. 1.17a in that $r(T)$ includes *all* recombination processes. Therefore, it depends not only on temperature, but also on the presence of imperfections in the crystal that act as recombination centers. In particular $r(T)$ may differ from one sample to another at the same temperature, although it does *not* depend on n and p directly (provided they remain small enough to validate Eq. 1.39). That is, r certainly does not depend on n' and p' (subject to the aforementioned size limitation on n and p), and it also does not depend on n_o or p_o *directly*, either.

But it may happen that altering the doping of a semiconductor not only changes the equilibrium carrier concentrations, but also changes the crystal perfection. Indeed this is quite commonly true, so that r may differ in two samples of Ge at the same temperature, whether the samples have the *same* or *different* values of n_o and p_o [remember though that $n_o p_o = n_i^2(T)$]. This difference in r arises because of side effects, which acting by themselves would not have altered n_o and p_o, but which simply accompany the metallurgical conditions involved in preparing the material initially.

If we apply Eqs. 1.40 to Eq. 1.39, we find that

$$R(n, p, T) = r(T)n_o p_o + r(T)n_o p' + r(T)p_o n' + r(T)n'p' \tag{1.41}$$

As suggested above, we usually do not care about the *total* recombination rate $R(n, p, T)$, because part of it is cancelled by the *total* equilibrium thermal generation rate G_{th}. But we do care about the recombination rate for *excess* pairs; that is, the rate at which the system tries to re-establish thermal equilibrium. This rate is given by Eq. 1.41 as

$$R(n, p, T) - G_{th} = [r(T)n_o p_o - G_{th}] + r(T)n_o p' + r(T)p_o n' + r(T)n'p' \tag{1.42}$$

But we have seen a balance of generation and recombination for *one* process in thermal equilibrium in connection with Eq. 1.17a; and we have pointed out that each and every process *self-balances* under the equilibrium condition. So the *entire* equilibrium ($n' = p' = 0$) recombination rate, $r(T)n_o p_o$, must be cancelled by the *entire* equilibrium thermal generation rate G_{th},

$$G_{th} = r(T)n_o p_o \tag{1.43}$$

Thus Eq. 1.42 becomes:

$$R(n, p, T) - G_{th} = r(T)n_o p' + r(T)p_o n' + r(T)n'p' \tag{1.44}$$

This last result is, unfortunately, nonlinear in the *excess* concentrations n' and p'; but there are two circumstances in which it may be simplified considerably. One is rather special. It occurs if *both* excess concentrations are much smaller than their *respective* equilibrium concentrations. Then

$$R(n, p, T) - G_{th} \simeq r(T)[n_o p' + p_o n']; \qquad \text{if } n' \ll n_o \textit{ and } p' \ll p_o \tag{1.45}$$

which is a linear relationship in the excess concentrations. It says simply that, when only a few excess carriers are added, the *added* ones do not recombine often with *each other*, but instead do so with the opposite species of carriers already there.

The other circumstance in which Eq. 1.44 may be simplified is much more important. It occurs if the material is dominantly either p-type or n-type ($p_o \gg n_o$ or $n_o \gg p_o$), and the excess carrier concentrations are comparable to each other and both are much smaller than the *majority*-carrier concentration. For the n-type situation of this type, Eq. 1.42 becomes

$$[R(n, p, T) - G_{th}]_{n\text{-type}} = r(T)n_o p'; \qquad \text{if } n_o \gg p_o \textit{ and } n' \sim p' \ll n_o \tag{1.46}$$

and for the p-type case

$$[R(n, p, T) - G_{th}]_{p\text{-type}} = r(T)p_o n'; \qquad \text{if } p_o \gg n_o \textit{ and } p' \sim n' \ll p_o \tag{1.47}$$

These results say that in the presence of a very large background of one (majority) carrier type, the relatively small number of excess carriers do not recombine often with each other, nor will the majority-type member of the pair find a background *minority* carrier with which to recombine. Instead, the major number of recombinations will occur because an *added minority*-type member of the pair easily finds an *equilibrium majority* carrier.

In the problem of uniform illumination that we started with, it is clear that $p' = n'$ everywhere, and there is no difficulty about the condition that these excess concentrations be comparable, as required in Eqs. 1.46 and 1.47. In other problems, this relationship will not be obvious; but we shall see in Chapter 4 that it is nevertheless almost always true in the dominantly n-type or p-type material that we use for most semiconductor devices.

Notice that the conditions in Eqs. 1.46 and 1.47 do *not* require that the excess minority-carrier concentration be small compared to the equilibrium minority-carrier concentration. Not at all. *We can indeed have a very large percentage change in minority-carrier concentration, and still use Eqs. 1.46 and 1.47.* Thus the reason that more than just arbitrary convenience attaches to describing excess-carrier processes as *minority-carrier* injection and recombination stems from the appearance of only the *excess minority* carrier in Eqs. 1.46 and 1.47. In fact, it is usual to rewrite these relations in the form

$$\left[-\frac{d\eta}{dt}\right]_{\text{recomb.}} = \frac{\eta'}{\tau_\eta} = \frac{(\eta - \eta_o)}{\tau_\eta} \tag{1.48}$$

where η represents the *minority*-carrier concentration and the parameter τ_η is called the *minority-carrier lifetime*. Representative values of τ_η range from 0.1 to 1000 μsecs and depend on the structure and doping [$r(T)$ and n_o or p_o] of the semiconductor. For the uniform-bar situation we started with, the solution to Eq. 1.48 after the light pulse is:

$$\eta' = \eta'(0)e^{-t/\tau_\eta} \tag{1.49}$$

where $\eta'(0)$ is the excess minority-carrier concentration at $t = 0$, just after the light pulse.

1.6.3 *Diffusion*

Now suppose we perform our light-flash experiment on the uniform bar in a different way. Instead of having the light flood the whole length of the bar, let it be confined to a narrow segment near the middle. Imagine that the bar is very long, to eliminate worry about the ends.

Once again, after the flash a neutral cloud of holes and free electrons is formed, but now only over the center section. Of course, this cloud will collapse by recombination, just as before—but with a new feature! The concentration of excess carriers during the recombination process is larger near the middle of the bar than at some distance away from the middle.

Now, if the injection is not too large, so we do not throw the system very far out of equilibrium anywhere, the injected carriers share the thermal motion of the lattice, as do all the carriers. Even though the motion is completely random, in the sense that each particle has an equal probability of going either to the right or to the left after each collision with the lattice, there will be more particles moving outward from the center than moving inward from the ends (P1.21 and P3.3)! The probabilities *per particle* may be the same everywhere, but there are plainly *more particles* near the middle than near the ends. Hence, there will be a net flux of particles outward from the center, by a process known as *diffusion*. It is the same process that spreads a drop of ink horizontally throughout a still pan of water.

There are several attributes of this diffusion flow that require emphasis:

First, there is no net force on *any* particle in the direction of flow.

Second, the particles do not have to push each other from the rear, like the crowd behind a police line. Indeed there need be *no* collisions at all between the diffusing particles to "make them go."

Third, the diffusing particles *must* be agitated at random by, and collide with, the surrounding medium in which they are immersed, like balls on a level, horizontally shaking pin-ball machine, so that each one has an equal likelihood of going in any direction after each collision.

Finally, although each particle has an even chance of going in any direction between its collisions with its surroundings, the key point is that the chance of its ever returning to a position it held "several collisions earlier" is practically nill. The particles are always basically "leaving every place they have ever been."

It is clear that the flux of diffusion depends on a difference of the concentration of the particles involved, at adjacent points in space. Mathematically we say that we have a *gradient* of the concentration $\nabla\eta$ of the minority carrier, and to the first order the flux is proportional to this gradient:

$$\mathbf{J}_{\text{particle}} = -D_\eta \nabla\eta \tag{1.50}$$

Here, D_η is by definition independent of $\nabla\eta$; but it may depend on η and, of course, on ambient conditions like temperature T. In most cases of interest to us, however, D_η will not depend on η (the concentration of the diffusing minority carrier), but it may depend on the temperature and on the background *majority*-carrier population (i.e., through the scattering effect of charged impurity atoms in the lattice).

The units of $\mathbf{J}_{\text{particle}}$ are "number of particles per unit area, per sec." It is a particle flux density. But simply multiplying it by the electric charge per carrier converts it to a current density.

The parameter D_η in Eq. 1.50 is called the *diffusion coefficient*, and it has dimensions of (length)^2/time. The negative sign accounts for the fact that diffusion flux proceeds *from* the region of high concentration, *to* the region of low concentration, whereas the concentration gradient, by definition, points the other way. Care must be taken with algebraic signs when finding the electric currents carried by diffusing positive and negative charges.

The reason why diffusion becomes important in semiconductor problems like the present one involving optical injection, and not usually in connection with metallic conduction, hinges on the electrical neutrality of the mobile cloud. In the case of reasonably heavily doped material, the majority carriers are plentiful and mobile. They can move in to "screen" the minority carriers from exerting on each other the mutual repulsion that occurs in a system with only one type of *mobile* carrier (P1.18). Because of the possibility of these screening effects, the flow of minority-carrier current by diffusion is extremely important in semiconductor devices. For these minority carriers, it is probably even more important in electronic devices than the current flow associated with the electric field.

PROBLEMS

P1.1 Considering the lattices of Fig. 1.1, estimate the number of atoms/cm^3 of a solid having each one for its crystalline form.

P1.2 The actual number of atoms N/cm^3 of any substance can be calculated from Avogadro's number N_o ($= 6.02 \times 10^{23}$ molecules per gram-mole), the atomic weight A of the substance, and its density ρ. Look up A and ρ for the following substances, and find N: (*a*) crystalline carbon (diamond); (*b*) silicon (Si); (*c*) germanium (Ge); (*d*) platinum (Pt); (*e*) mercury (Hg); (*f*) potassium (K); (*g*) sulfur (S); (*h*) any monatomic gas at standard pressure and temperature. Compare the results with that of P1.1. What do your results suggest about the dependence of interatomic forces upon distance?

P1.3 Use the data of Fig. 1.4 to determine the number of atoms/cm³ of C, Ge, and Si. Compare with the results of P1.2.

P1.4 Assuming that the atoms of the lattice share the kinetic energy of thermal motion as if they were free, find out how well quantum mechanics indicates that they are localized.

P1.5 Using Eqs. 1.10 and 1.11, work out the ionization energy of hydrogen. Express the result in electron volts.

P1.6 (*a*) By how much does the mass of a semiconductor change when a hole-electron pair is produced? When it recombines?

(*b*) Repeat the development of Eq. 1.15 for holes, with due attention to algebraic sign. Do the same for Eq. 1.27.

(*c*) Strictly speaking, it is not obvious that the *effective* mass m^* of a carrier of charge q will be a positive number. Suppose we have the problem of determining experimentally the sign of q and the sign of m^* for some unknown carrier in a solid. Discuss how a study of the currents and voltages resulting from electric and magnetic fields, applied either separately or simultaneously, can unambiguously distinguish between all possible combinations of sign. Concern yourself also with the algebraic sign of τ_c, and give evidence to determine it.

(*d*) Show that the Hall effect for *p*-type material would not have the polarity which is actually observed for it, if the motion of holes were really just a retrograde motion of electrons.

(*e*) Consider a very long cylinder of a homogeneous extrinsic semiconductor of radius R_o, spinning about its longitudinal axis at a steady angular speed ω_0 rad/sec. Neglecting end-effects and diffusion, find the steady-state electric field inside and outside the cylinder. Find also the net volume charge density and surface charge density on the cylinder, per unit length. Do this for both *n*-type and *p*-type material. Also discuss the current flow.

P1.7 In connection with mobility, we are interested in estimating the (small) energy loss of an electron per collision with the lattice. A similar problem is the following. Considering only motions along one dimension, show that in perfectly elastic collisions between a light and a heavy particle energy will, on the average over all possible initial circumstances, transfer from the more to the less energetic particle in approximately the ratio of four times the smaller to the larger mass.

P1.8 Estimate very crudely the temperature dependence of μ for collisions with the vibrating atoms as follows. Argue that the average potential energy of a vibrating atom is proportional to kT, so it spreads itself out in a sphere of effective cross section proportional to T. Argue next that the number of collisions an electron will make in moving a unit *distance* is proportional to the cross section found above, so the mean distance (mean free path) between collisions is proportional to T^{-1}. Finally, take two cases: first, argue that the average kinetic energy of the free electrons is $3kT/2$, in which case $\mu \sim T^{-3/2}$; second, take the case in which the kinetic energy of the free electrons is independent of temperature, which leads to $\mu \sim T^{-1}$. It turns out that the first case corresponds very approximately to a semiconductor (but see Table 1.2 and Fig. 1.7), and the second to a metal.

P1.9 (*a*) If a constant τ_{th} is the mean free time of all electrons for collisions with the lattice atoms alone, and a constant τ_i is the mean free time for collisions with charged donor impurities present in a given concentration alone, argue from the probability of independent events that τ_c for both processes together is given by $1/\tau_c = 1/\tau_{th} + 1/\tau_i$. How does τ_i depend on the concentration of ionized donors N_d^+?

(*b*) Use these ideas to complete the following table, and explain any other assumptions you make.

Si Sample (300°K)	μ_e(cm²/v-sec)	μ_h	σ mhos/cm
1 Intrinsic:	1350	480	?
2 Doped with 10^{16} atoms/cm³ of P:	1100	250	?
3 Doped with 10^{15} atoms/cm³ of P:	?	?	?
4 Doped with 10^{16} atoms/cm³ of B:	?	?	?
5 Doped with 10^{16} atoms/cm³ of B *and* 10^{16} atoms/cm³ of P:	?	?	?

Compare your results with Fig. 1.8. Comment on why in that figure the mobility of a given type of carrier is lower when it is a minority carrier than when it is a majority carrier.

P1.10 Work out some numerical values of $<\tau_c>$ for Ge and Si from the tables and graphs given in the text. Become familiar with the orders of magnitude. How far does light travel in time $<\tau_c>$?

P1.11 What percentage of covalent bonds is broken in pure Ge at 300°K? In pure Si at 300°K? Repeat for a temperature of 400°K. Repeat at both temperatures if donors are added to both materials in the concentration $10^{14}/cm^3$. Repeat for acceptors in the same concentration.

P1.12 The equilibrium concentrations of holes and conduction electrons in a homogeneous semiconductor are determined exactly by Eqs. 1.18 and 1.20, provided we know that all the impurities are ionized, and are given the value of n_i at the temperature involved.

(*a*) Derive exact expressions for n_o and p_o in terms of N and n_i.

(*b*) For n-type material, with $N > 0$, at what values of N/n_i is n_o within 5% of N? At such values of N/n_i, compute p_o/n_o.

(*c*) Using the results of (*b*), determine the minimum value of N in Ge at 300°K for which $n_o \cong N$ and $p_o \cong n_i^2/N$ are correct to 5%. Calculate the fraction of Ge atoms replaced by donors to reach this doping level (if $N_a = 0$). Calculate also the resistivity of Ge at this doping level.

(*d*) Repeat part (*c*) for Si at 300°K.

P1.13 Find p_o and n_o at 300°K in Ge containing 2.4×10^{13} atoms/cm³ of Sb. Repeat at 200°K and 400°K.

P1.14 Suppose 4.8×10^{13} atoms/cm³ of In are added to the Ge of P1.13, *in addition* to the Sb. Answer the same questions asked there. Repeat if the In is added *instead* of the Sb.

P1.15 (*a*) Estimate the mobility for a typical metal having a resistivity of 10^{-5} ohm-cm.

(*b*) Estimate the mean free time $\langle\tau_c\rangle$, if $m^* = m$.

P1.16 Using the tables, graphs, and equations referred to in the text, find the intrinsic conductivity of germanium and silicon at temperatures of 100, 200, 300, 400, and 500°K. Discuss any extrapolations of the given data you have to make.

P1.17 Assuming that the mobilities are independent of impurity concentration, find the minimum conductivity of a given semiconductor material, achievable by choice of impurity concentration, and the conditions under which it occurs. Compare the minimum with the intrinsic conductivity for Ge and Si at 300°K.

P1.18 Consider a homogeneous one-carrier conductor of conductivity σ and dielectric permittivity ϵ. Imagine a given distribution of the mobile charge density $\rho(x, y, z; t = 0)$ in space, at $t = 0$. We know the following facts from electromagnetism, provided we neglect diffusion current:

$$\nabla \cdot \mathbf{D} = \rho; \qquad \mathbf{D} = \epsilon\mathbf{E}; \qquad \mathbf{J} = \sigma\mathbf{E}; \qquad \nabla \cdot \mathbf{J} = -d\rho/dt$$

Show from these facts that $\rho(x, y, z; t) = \rho(x, y, z; t = 0)e^{-t/(\epsilon/\sigma)}$. Interpret this result to show that mobile charge cannot remain in the bulk of uniform conducting material, but must accumulate at surfaces of discontinuity or other places of nonuniformity. Compute the value of the "dielectric relaxation time" ϵ/σ for a typical metal, and for various extreme and ordinary dopings of Si and Ge.

P1.19 (*a*) Derive the Hall coefficient R at moderate magnetic fields for a two-carrier semiconductor, in which the mean free times are independent of energy (see Laboratory A2.4).

(*b*) Sketch $R(T)$ versus T for a p-type sample. Assume that $\mu_e/\mu_h \equiv b > 1$ and that b is independent of T. Find the minimum value of $R(T)$. Compare the sketch with one for $R(T)$ for an n-type sample under the same conditions.

P1.20 For a *one*-carrier semiconductor, find the Hall coefficient R for moderate magnetic fields, if the mean free time *is* a function of the kinetic energy. Express the result in terms of various "averages" of the mean free time.

P1.21 Consider a simple model for diffusion in one space dimension. Assume that all particles must move only at discrete regularly spaced times, one time unit apart. When they move, each particle may only jump one space unit either to the left or to the right, with *equal* probability. Start with 1024 particles at $x = 0$, $t = 0$, and build up step-by-step the pattern of particles in space after 10 time units. Plot $\ln_2 n$ *vs* x^2 or $\ln_2 x^2$ in space after each jump and measure the corresponding width W between points of one-half maximum concentration. Then plot W^2 versus time. What does your result suggest about the rate of "spread" of particles by the diffusion process (in *one* space dimension, at least)?

2

The Energy-Band Model of a Semiconductor

2.0 INTRODUCTION

In our previous considerations we have introduced the distinction between metals, semiconductors, and insulators through a primarily qualitative discussion which has emphasized the location in space of bound electrons, conduction electrons, holes, donors, and acceptors. All of this constitutes what is usually referred to simply as the *bond* model of a semiconductor. As a matter of fact, we have had to point out several times that, because of the underlying quantum nature of the problem, the pictures we have used tend to be overlocalized in character and to suggest details of the dynamics of electrons in solids that are not valid. We have introduced some elementary wave mechanics to counteract some of the misimpressions that may otherwise have been implied.

Various energies have been parameters in the discussion. For example, the "ionization" energy E_g required to break a valence bond has played an important part, as have the binding energies of extra carriers to donors and acceptors in the lattice, and the thermal energy possessed by a carrier that is in thermal equilibrium with the surrounding lattice.

There is another more quantitative viewpoint in terms of which we can discuss semiconductors. It is much more closely related to energy and momentum concepts than it is to the space-location and velocity concepts that we have so far been using. Essentially quantum mechanical in nature, it is known as the *energy-band* viewpoint, and it is often helpful enough in connection with discussions of semiconductor devices to justify our consideration here. In spite of the fact that we cannot now develop it completely upon its own foundations, as we really should, we do wish to translate with some care our previous picture into the new one.

2.1 ATOMIC STATES AND ENERGY LEVELS

To make the transition from the bond model to what is called the *band* model, we must discuss the behavior of solids (metals, insulators, and semiconductors) in terms of the same elementary quantum notions we use to describe isolated atoms.

We are already familiar with the idea that the various atomic electrons are permitted to adopt orbits, or modes of motion, with prescribed angular momenta. These orbital angular momenta, as well as the values of the spin of the electron itself, are selected by quantum conditions. The corresponding modes of motion or states are, in the absence of disturbing forces, a set of discrete entities that arise from the wave nature of the electron. In general, there is associated with each such state a particular value of the energy of the particle, although often more than one state has the same value of energy.

Thus we employ "energy-level diagrams" which describe the states that an electron in an atom is permitted to have, similar to the diagram shown in Fig. 2.1. The periodic table, Fig. 1.3, is associated with the fact that the various states lie in shells which are only able to accommodate particular numbers of electrons, these numbers increasing with the energy corresponding to the shell. The principle that leads to this limitation on number is the *Pauli exclusion principle*, which says essentially that no two electrons in the same system can have the same state. Included in the description of the state of an electron in an atom are its total energy, magnitude of orbital angular momentum, and components of both the orbital and the electron spin angular momenta along

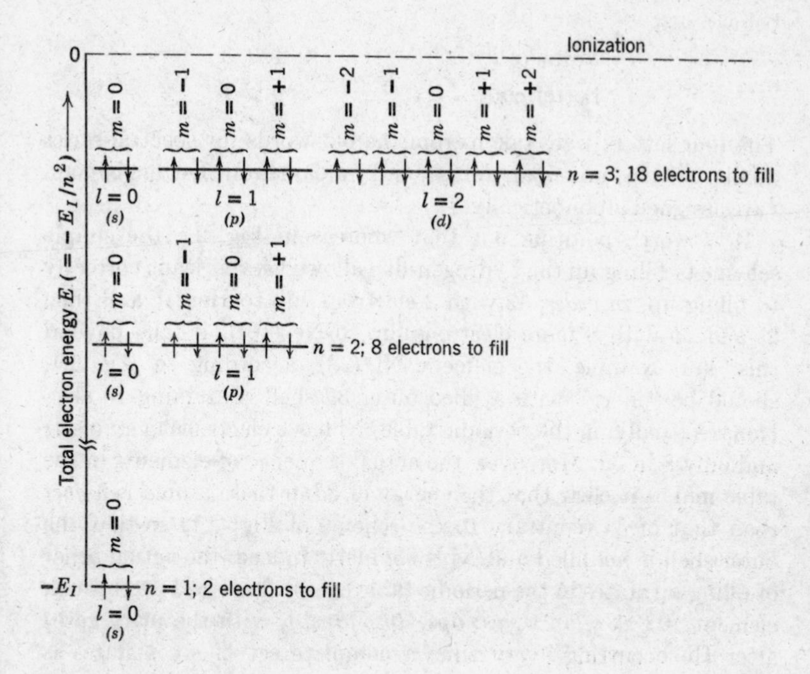

Fig. 2.1. Representation of some of the allowed states and energy levels for the single electron in a hydrogen atom. The number n identifies the total electron energy, l the magnitude of its orbital angular momentum, and m the axial component of that momentum. If electrons in multi-electron atoms did not affect each other any more than just to neutralize part of the nuclear charge, they would have a very similar arrangement of allowed states, and fit into them as illustrated here (the arrows show the electron spin). Reference to Fig. 1.3 shows, however, that the detailed pattern of relative energies of the states implied here breaks down beyond Ne (10).

the space axis of reference chosen for the description of the atom. Thus there are four numbers required to describe the state of each electron in an atom, and no two electrons in the atom are permitted to have the same four "quantum numbers."

In the periodic table reproduced in Fig. 1.3, the information on the outer-electron configuration of the atoms is given in the standard code of spectroscopy. This code gives the shell first, which is the value of n, and then gives the orbital momentum by designating the values of l with a letter equivalent, according to the scheme below:

Value of l	0 1 2 3
Letter code.............	s p d f

The four letters were taken from the old words for spectral-series quality: *s*harp, *p*rincipal, *d*iffuse, and *f*undamental. Letters beyond *f* are assigned alphabetically.

It is worth pointing out that whereas in Fig. 2.1 the simple scheme of filling up the hydrogen-like allowed states leads correctly to filling up, *in order*, $1s$ with 2 electrons [up to He(2)], and then $2s$ and $2p$ with 8 more electrons [up to Ne (10)], it fails beyond this. For example, the element Ni (28), according to Fig. 2.1, should be "inert" with a filled outer $3d$ shell containing 10 electrons. Actually, in the periodic table, Ni has 2 electrons in states $4s$ and only 8 in $3d$. Moreover, the actual sequence of elements in the table makes it clear that the energy of $3d$ in these atoms is *higher* than that of $4s$ (contrary to the scheme of Fig. 2.1) so that the outer shell is *not* filled and Ni is *not* inert. Indeed, the actual order of filling up states in the periodic table can be described up through element 102 as *s, s; ps, ps; dps, dps; fdps, f*, with the inert gases after He occurring every time a complete set of six *p*-states is filled with electrons. This is contrasted with the scheme *s; sp; spd; spdf; spdfg* implied by Fig. 2.1, in which "inert" elements would occur only with filled *s, p, d,* and *f* shells successively. A closer study of the quantum mechanics of atoms would be needed to explain these detailed features of the periodic table, as well as why the values of l and m are restricted to the ranges $0 \leq l \leq (n-1)$ and $-l \leq m \leq +l$.

If an electron makes a transition from one state to another in response to some external force, the *discrete* difference in energy between the two states must be accounted for through a change in the electromagnetic or mechanical energy of the surroundings. This may be effected by emission or absorption of a photon of electromagnetic radiation, or, in the case of an atom in a crystal, by changes in the mechanical vibration energy of the rest of the crystal. According to the principles of quantum mechanics, the amplitude of the mechanical vibration of the lattice at any of its natural frequencies ν may only change by *discrete* amounts, such that the corresponding energy is a whole multiple of $h\nu$. Mechanical vibrations with energy equal to $h\nu$ are called *phonons*, and relate to sound waves in the same way that photons relate to light waves.

2.2 BAND STRUCTURE OF AN "INTRINSIC" SEMICONDUCTOR

2.2.1 *Development of Bands by Atomic Coupling*

Let us next consider a volume of crystalline material containing N atoms. We could imagine the N nuclei as being fixed in position in their periodic array, and ask what would be expected to be the permitted "orbits," or dynamical states, of all of the electrons required to neutralize these nuclei. In other words, we could look upon the section of crystal as a very large and complicated molecule, or multi-nucleus atom.

Just as is the case in a single atom, we would expect to find a multiplicity of allowed states for these electrons, each one representing in general some sort of orbital path through the crystal for an electron "occupying" it, and characterized by an individual value of the electron energy. It is pertinent to ask whether there are any general features of the resulting energy and state picture that can be anticipated without performing a detailed calculation for such a complicated system.

One guide to the energy-level structure for a crystalline system is furnished by imagining first what would happen if the spacing between the atoms were expanded to the point where all of them were extremely far apart. Under these circumstances, we would expect that the allowed states for the entire system of N atoms would just be the pattern of allowed states for one atom alone, repeated N times. That is to say, all the allowed states of the complete system could be obtained simply by considering every possible combination of the allowed states for the individual atoms, treated as separate and noninterfering units. The situation could be represented by Fig. 2.1, in which each horizontal line actually becomes N lines on top of one another, and the maximum number of electrons that could be accommodated on each of these "multiplet" lines would be N times the number shown on the figure. No matter what state the system as a whole might be in, there would be no possibility of an electron transferring itself from one atom to another when the atoms were so far apart; so each state of the system would correspond simply to a specification of how each of the electrons associated with each of the individual atoms was orbiting about its own nucleus.

If we imagine next that the atoms are brought somewhat closer together in the crystalline structure, it will become clear that the outermost electron of each atom (in its lowest state) will eventually come close enough to another atom to feel the forces exerted by the latter. At such a spacing, the innermost electrons on each atom will not actually have come close enough together to feel any disturbance from the neighbors, and thus the states for these inner electrons will be expected to remain approximately those of the isolated atoms repeated N times, as before. When the atomic spacing is small enough so the outermost electron orbits would begin to overlap if they remained unaltered by the proximity of atomic neighbors, these orbits would in fact be expected to become altered considerably. Indeed, electrons in these new orbits could very well exchange positions from one atom to the next, so that the resulting states would involve orbiting motions throughout the entire crystal.

One of the most important points in this situation is the fact that, if we consider two such atoms having two separate states that are identical at large interatomic separations, we must get two new states, possibly at different energies, after the atoms are brought closer together. The total number of degrees of freedom for the two electrons which are always involved in the problem is unaltered as the atoms are brought together, so the number of modes of dynamical motion also remains unaltered. Looked at in another way, we can see that the number of initial conditions independently specifiable for the dynamics of the two particles is always the same, regardless of whether the interference between them is large or small.

Of course, when N atoms are brought together in a similar fashion, we can expect the N identical motions, corresponding to the orbits for a particular dynamical motion of each electron about each atom separately, to become N new orbits, possibly with different energies, after the mutual interference or coupling begins to take place.

The situation described above, depending as it does only on the number of degrees of freedom of the dynamical systems involved, can also be illustrated in terms of an electrical problem which is perhaps somewhat more familiar to us than the atomic one under consideration. An appropriate case is that of two identical tuned *LC* circuits, capable of interacting with each other through mutual inductance between the coils. These two tuned circuits, whether coupled or not, always have a total of four degrees of freedom, corresponding to the fact that initial current could be specified independently in each of the two inductors, and initial voltage could be specified independently on each of the two capacitors. The resulting system is described by two *independent* second-order differential equations if there is no mutual coupling, and by either two *dependent* second-order differential equations, or one fourth-order differential equation, if there is coupling; but the total number of undetermined constants in the solution is always four, corresponding to the number of independently specifiable initial conditions. The number of degrees of freedom of the system is unaltered by the action of the mutual coupling.

In this circumstance, we know that without any coupling between the resonant circuits there are two identical natural frequencies for the system, corresponding to the fact that either circuit can be oscillating by itself at the same radian frequency $\omega_0 = 1/\sqrt{LC}$. When the circuits are coupled together, however, we know that, although there remain two natural frequencies for the system, the values of the frequency are no longer the same as for the isolated circuits alone. In fact, in this case one of the frequencies happens to be slightly above the isolated natural frequency and the other is slightly below; but the most important point is that the difference between the two frequencies increases with the strength of the coupling between the circuits. We can say that the two "states" which were identical without any interaction between the tuned circuits, are *split* by the interaction into two new "states" with different characteristic frequencies.

It is not difficult to see that with three or more identical circuits coupled together there would be a corresponding number of new natural frequencies, all differing from the one characteristic of the isolated circuits, so that for three tuned circuits mutually coupled we would expect three new frequencies; for four we should expect four new frequencies; etc. These new frequencies represent the characteristic oscillation frequencies for the coupled system, and when the system is oscillating at one of these frequencies alone there is a particular pattern, "mode," or "state" of currents and voltages in all of the tuned circuits that is characteristic of the particular frequency concerned.

For example, for two identical tuned circuits coupled together, there are two simple patterns of behavior corresponding to the two natural frequencies of the system. This situation is illustrated in Fig. 2.2, where it will be noted that in the mode of oscillation

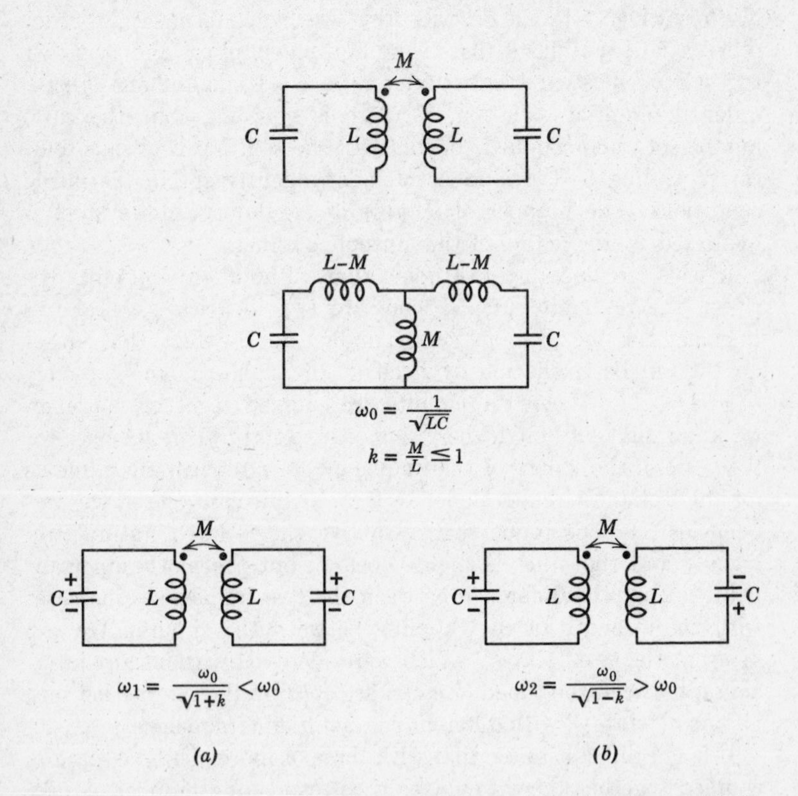

Fig. 2.2. Coupled tuned circuits, showing (*a*) symmetric mode of oscillation corresponding to a natural frequency ω_1 lower than the uncoupled frequency ω_0, and (*b*) antisymmetric mode with natural frequency ω_2 higher than the uncoupled frequency ω_0.

corresponding to the lower natural frequency ω_1, the two capacitor voltages are oscillating in phase with each other, whereas in the pattern of oscillation corresponding to the higher natural frequency ω_2, the voltages on the two capacitors are exactly out of phase with each other. In the latter case, no current is passing through the center arm of the T of inductances that represents the transformer. Thus, in the present circumstance the lower natural frequency corresponds to a mode of oscillation which is symmetrical in the voltage pattern in the circuit, whereas the higher natural frequency corresponds to a mode which is antisymmetrical in the voltage (P2.1).

In connection with the electrons in the N atoms that we were considering previously, the detailed situation will be more complex, but there will still be a lowest and a highest value in the new set of energies corresponding to the N new orbital states which have "split off" from the N identical old ones. We can expect the differences between the new and old energies to increase as the coupling between the originally isolated orbits gets closer.

Now, in terms of our previous bond picture, we have pointed out that, at the actual interatomic distance in semiconductor crystals such as Ge and Si, only the outer orbits, or valence-electron states, of the isolated atoms would overlap enough to alter their character significantly. The inner atomic electrons, which we have taken as "core" electrons in our previous discussions, would be expected to have orbital radii much smaller than those of the valence electrons in the isolated atoms (apply suitable modifications of Eqs. 1.5 to 1.12 to determine orbits for different nuclear charges). These inner electrons would then lead to *system* orbital states which are very nearly N times the corresponding states for isolated atoms. This situation simply corresponds to the fact that electrons which possess these "core" motions *are*, in fact, almost 100% localized on the atoms with which they were originally associated, and do not get shared among the various crystal atoms to any significant degree. What happens in more detail to the valence electron states, of which four per atom are occupied by electrons in the case of Ge and Si, is the more interesting question.

If we look back at Fig. 2.1, and also at the periodic table in Fig. 1.3, we find that the elements C, Si, Ge, Sn, and Pb, which make up the cubic part of Group IV, all have two electrons each in the orbits with momenta corresponding to *s*-states ($l = 0$) and *p*-states ($l = 1$); but the respective major energy shells in which they lie are $n = 2$ for C, $n = 3$ for Si, $n = 4$ for Ge, $n = 5$ for Sn, and $n = 6$ for Pb. Moreover, Si and Ge crystallize in the same structure as C (e.g., the *diamond* structure), and on this basis it is no surprise that the properties of these three crystals are so nearly scalable one from the other.

Considering C, Si, and Ge, then, the states for the valence shell and the next lower one are shown in Fig. 2.3 as a function of the spacing of the atoms in the crystal, assuming that we could vary that spacing at will (which, of course, we cannot). At large spacing, we see from Fig. 2.1 that the states with lowest orbital momentum in the outermost shell for these materials have states enough for 2 electrons in $l = 0$ and 6 electrons in $l = 1$. Hence in Fig. 2.3, which applies to N atoms, the number of available states for $l = 1$ is $6N$, and for $l = 0$ it is $2N$.

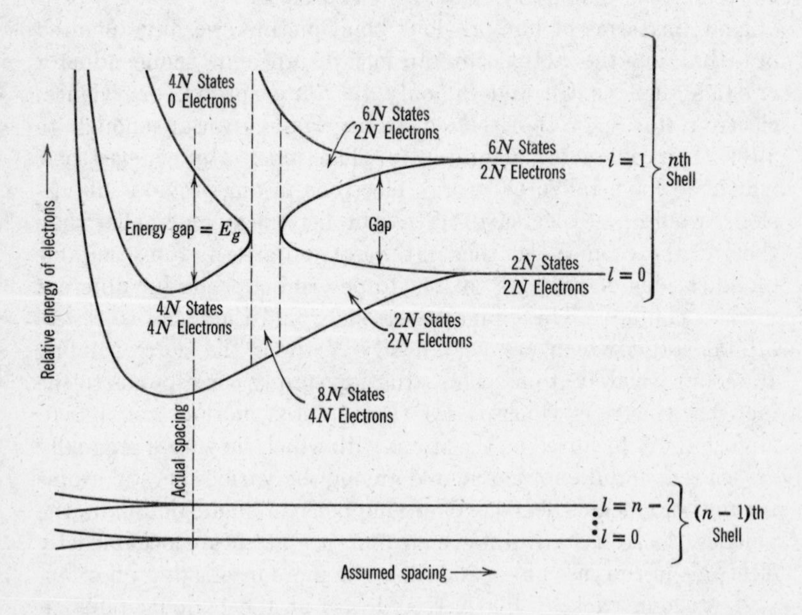

Fig. 2.3. Energy levels in the diamond structure, as a function of assumed atomic spacing (theoretical; calculation carried out for carbon).

Note, however, from the periodic table, that in the materials we are considering, the $l = 0$ (or s) states are fully occupied by 2 electrons in each isolated atom, or $2N$ electrons in the crystal; but the six available $l = 1$ (or p) states per atom are only one-third occupied by the two remaining valence electrons per atom, or $2N$ such electrons in the crystal.

As the spacing of the atoms is imagined to decrease, the $6N$ states for $l = 1$, which originally had identical energies in the isolated atoms, become distinct instead and form an *energy band.* Similarly, the $2N$ states for $l = 0$, with identical energies in the isolated atoms, split into a second band, separated from the first one by an *energy gap.* In Fig. 2.3 we have shown the upper and lower energy levels in each band. Because the amount of splitting of identical levels (e.g., the energy difference between the lowest and highest states in the band) is a measure of the amount of mutual interaction, whereas the original spacing between the $l = 0$ and $l = 1$ atomic levels is set by the original atomic structure, it is clear that for large but finite lattice spacings the bands must be much narrower than the gap.

At still smaller spacings, however, there comes a region in which the bands merge with each other. The lowest energy state that splits from the upper atomic level appears to drop below the upper state that has come from the lower atomic level. When this happens, there is simply a region of new dynamical behavior that is a mixture of both the upper *and* lower original states. This region cannot be regarded as closer in type to one state than to the other. There exists no energy gap, and there is just one band. In the present case, that band has twice as many available states as there are available electrons in the material.

Finally, at the "actual spacing," the bands have again split apart, separated by the "energy gap E_g," and the resulting motions of the electrons corresponding to the new states are quite mixed in character, between both of the original ones from which they came. Indeed, the number of states has been reapportioned between the two bands, although there are still $8N$ states in all. The significant point is that there now happens to be exactly as many states in the lower band ($4N$) as there are available valence electrons from the atoms ($4N$), so that this band can be completely filled while the upper band is completely empty.

Another study of Fig. 2.3 shows that the energy of the electrons can be reduced by making the atomic spacing somewhat closer than that indicated as the "actual spacing." Such a reduction of energy would therefore seem to increase the binding energy of the atoms, inasmuch as it would have to be supplied to them again if they were isolated. The binding energy of the electrons alone would indeed be greatest at the minimum energy point of the curves. The charged nuclei, however, repel one another. Thus, for the entire atom to have minimum energy, the actual spacing has to be to the right of the one that yields minimum energy for the electrons alone.

Bear in mind that there is no *simple* reason why the system states should or should not behave in the particular detailed way shown in Fig. 2.3. Theoretical prediction by quantum-mechanical methods of the results for any given material is quite difficult, even in the simplest cases, and has been carried out for only a few examples. Our interest here is first in the fact that energy bands separated by energy gaps certainly *can* occur for a system composed of many atoms.† Second, we call attention to the fact that when the bands are packed densely with the available number of electrons from the constituent atoms, the band of highest energy that contains any electrons at all can either be partly full or completely full.

Bands which arise from atomic levels normally occupied by electrons inside the valence or outer shell will be expected to be very narrow, as shown in Fig. 2.3, because the lattice spacing is relatively large for these. Thus, if these levels were completely closed (fully occupied) in the isolated atoms, we would expect the resulting bands to be filled also.

We may well inquire now about what happens to bands that arise from excited states of the isolated atoms. These states would lie above those shown at the extreme right in Fig. 2.3. In the atomic "ground state," no electrons would have such energies. Clearly these states correspond to even larger orbits than those we have been considering, and electrons in them would interfere with each other even more than the normal valence-state electrons do. Thus the splitting of these levels into bands would most likely occur

† Refer to the film described in the footnote on p. 59, to P2.4 and P3.2, and to Laboratory Experiment A3.

to a much larger extent than those heretofore considered, and it would seem that these bands would surely cross into the upper one of those shown in the figure by the time the actual lattice spacing was achieved. In other words, we do not expect any more energy gaps to occur above the one labeled E_g in Fig. 2.3. In general, this proves to be the case.

2.2.2 *Interpretation of the Bands*

We arrive finally at a picture of the electronic energy-band structure of a crystal with its normal lattice spacing. Considering the temperature to be at absolute zero, so that all electrons and atoms have their lowest possible energies, there are two cases to be distinguished, shown in Fig. 2.4. The type of representation shown is referred to as an *energy-band diagram*. In Fig. 2.4*a* appears the case of a metal, in which the highest band is only partly filled by electrons. This occurs, for example, in monovalent metals, where the band of energies arising from levels within the outermost energy shell contains two states per atom in the crystal

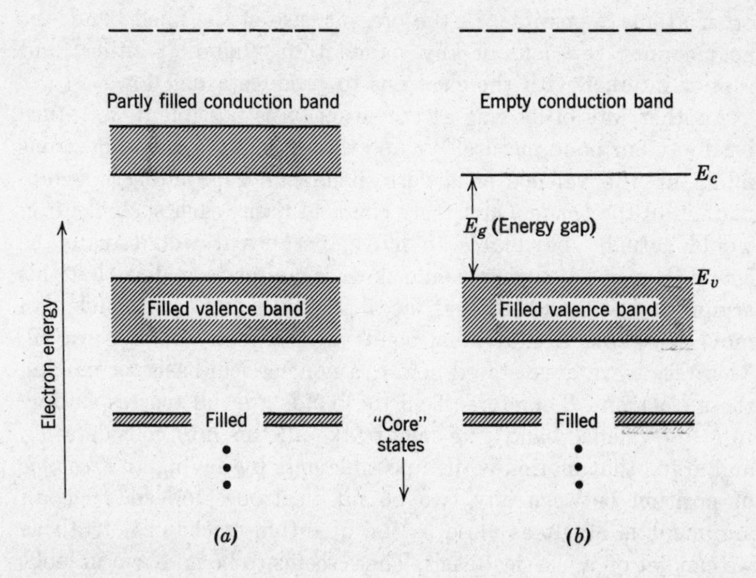

Fig. 2.4. Electronic energy bands for (*a*) metallic conductor at $T = 0°K$; (*b*) insulator or intrinsic semiconductor at $T = 0°K$.

(including electron spin), with only one electron per atom requiring accommodation.

In Fig. 2.4*b* is shown the other case, in which valence-electron states couple to give rise to two bands, separated by an energy gap E_g. The lower band has just enough states in it to accommodate the number of atomic valence electrons supplied by the component atoms. At low temperature, this lower or *valence band* of energies is just filled by the valence electrons, leaving the next higher band, *the conduction band*, completely empty. The significance of the filled band must now be considered.

To obtain an electric current from a small applied field, a number of electrons in a solid must be given some net momentum in the direction of the field. On the basis of quantum mechanics, however, *it is not possible to impart any such net momentum to a completely filled band*. One way of elaborating upon this result is to point out that there is some analogy here with the situation in a completely filled atomic shell, in which it is found that the electrons have no net angular momentum and therefore do not provide any momentum "handle" through which chemical interaction with other atoms is possible. In the present case of the filled band, the field cannot get hold of any momentum "handle" either, and cannot interact with the electrons to produce a net flow.

Another way of looking at the situation is possible if we return briefly to our bond picture. We find that the action of the electrons filling up the valence *band* corresponds to their filling up completely all the *bonds*. On a truly classical basis, each such electron would actually be unable to move, because it would really be *bound*. Hence no current would flow. We already realize that this argument is wrong, however, because of the fact that a hole can move; electrons in individual bonds are not really "stuck" at all. There is, as we have stated before, a nonclassical basis for moving these electrons. Therefore, if all the bonds are full (corresponding to a full valence band), we might take all this into consideration and argue that motion is then possible *only* by having an *exchange* of position between any two bound electrons. Indeed, such an argument is about as close to the quantum-mechanical truth as we can get on a classical basis. There seems to be no harm in looking at the situation this way, especially because the identical nature of all electrons means that such exchanges cannot really be detected anyway!

It is thus fair to take the position that, upon application of a field, a full band (or full bonds) cannot carry a net current, because on the basis of exchanges as the only permitted motion, as many electrons would move in one direction as in another.

Of course, when the band or bonds are not full, the exchange constraint is relieved and a net motion of the bound or valence-band electrons becomes possible. The resulting mode of motion is described most conveniently by the hole, because the details of the corresponding electron motions are complicated in the same sense that the motion of a liquid is complicated when a gas bubble rises in a glass of soda.

We are now in a position to appreciate why a partly filled top band, of the type illustrated in Fig. 2.4*a*, represents a conductor. In such a band, the electron motions are not constrained to exchanges; they can move quite freely. Hence, the material is expected to be essentially metallic in its electrical conduction properties. The freedom of carrier motion is available right down to extremely low temperatures.

In contrast with the metallic case in Fig. 2.4*a*, that of the insulator or semiconductor in Fig. 2.4*b* does not permit any electron motion at all at low temperature (near 0°K), because of the filled valence band and the empty conduction band. We would conclude that the material is an insulator, at least at the temperature involved.

If the temperature is raised, however, we expect some electrons to receive enough energy from the thermal vibrations of the crystal lattice to give these electrons at least the gap energy E_g. This means that a valence bond is broken, or, in the band picture, that an electron is raised from its energy in the valence band to one in the conduction band. The significance of the gap is that energy less than E_g will not suffice to remove an electron from the valence band at all. Energies greater than E_g suffice to move electrons in energy from below the top of the valence band E_v, or place them higher up in the conduction band than its bottom, E_c, or both. The major issue at the moment is the fact that raising an electron in this way leaves *both* the valence and the conduction bands *partly filled*. Therefore, electrons can carry current in *both* bands—those in the conduction band behaving like "free" electrons, and those in the valence band behaving in a manner best described by the conduction properties of a hole.

Hence the material represented by the electronic energy-band structure of Fig. 2.4*b* is essentially an insulator at very low temperature, but becomes more nearly a good conductor at higher temperatures. For, as the temperature rises, two kinds of carriers are generated in it in increasing numbers. Evidently, in such cases the distinction between an insulator and a semiconductor is more one of quantity than quality; the question is whether at a particular temperature of interest (room temperature, for example) *enough* electrons have been lifted from the valence band to the conduction band to call the material "conducting" at that temperature. If not, it would be classed as an insulator; if so, it would be called a semiconductor.

Accordingly, we may say that an intrinsic semiconductor is generally an insulator at low temperatures and a conductor at high ones, its conductivity tending to increase with temperature. We shall see in Chapter 3 that this occurs because in the intrinsic semiconductor the increase in the number of free carriers with temperature far outweighs the reduction of mobility.

The critical dimension in the problem of the generation of free carriers in the insulator or semiconductor is the size of the energy gap E_g compared to the thermal energy kT. At room temperature, $kT = 26$ millielectron volts. The energy gap in diamond is about 5.5 ev, and we have seen (Table 1.1) that in Si and Ge there are gaps of approximately 1.1 ev and 0.7 ev, respectively. At room temperature, diamond is certainly an insulator, with a resistivity of the order of 10^{18} ohm-cm. Silicon, when extremely pure, could in principle have a resistivity as high as 2×10^5 ohm-cm, but in practice such levels are achieved only by balancing donors against acceptors. Very pure germanium has the moderately high resistivity of 47 ohm-cm.

We have implied that at all normal temperatures the "core"-state energy bands will remain filled. The energy gaps below the valence band are, in general, large enough to assure this condition.

In regard to pictures of the type represented by Fig. 2.4, there is an implication that we have not yet stressed. It is connected with the meaning of the horizontal dimension, the axis of which has not been labeled. There will be reason later to associate distance in some space direction with this axis, and the sense in which this association is possible requires discussion. Strictly speaking, the energy-band diagram itself does not tell us any more than the possible energies of the various dynamical states that an electron in the corresponding system is *permitted* to have. As is usually done for individual atoms, the diagram is constructed (by making rather severe approximations) in such a way that, when the system contains many electrons, its total state is specified merely by giving the distribution of the electrons among the various "one-electron" states shown in the diagram.

It will be recalled that our development of the band picture of the intrinsic semiconductor was based on the coupling of *identical* atoms, and that the arrangement of these atoms in space had a repetitive, periodic, rule of formation corresponding to the crystal structure. If this periodicity of the structure were not present, the energy-band diagram would change with the size of the sample of the crystal considered. Each new atom added to the sample would introduce new couplings and modify old ones, just as adding new atoms to a molecule changes its properties significantly. The diagram would not be representative of the material alone, but also would involve its dimensions. For the periodic structure, however, the energy-band diagram is actually characteristic of any number of periods sliced out by imaginary boundaries in an otherwise infinitely repeated system. The number of states in any band does, of course, correspond to the number of atoms included within the boundaries selected, in accord with preserving the number of degrees of freedom; but such features as the energy gap magnitude and the width in energy of the various energy bands are independent of the number of periods included in the sample. This latter fact is true because in an infinite periodic array, the appearance of the whole system from *corresponding* points in any one of its periods is always the same. Thus, for example, the energy required to break a bond at any of the corresponding atoms in the structure is the same.

It is true that, for a real (finite) piece of material in the laboratory, the presence of boundary surfaces will make some basic lattice periods near these surfaces differ from others well inside; but normally the sizes of pieces we have to consider on a practical laboratory scale are large enough to permit us to treat the "inside" as if the surfaces were infinitely far away, and then to give independent attention to what happens just inside and outside the bounding surface.

Aside from the special surface problems already mentioned, to be handled separately in any case, the crucial issue at hand is the correlation, if any, between energy and space location. The point is that knowing an electron has any particular energy shown in the "band diagrams" of Fig. 2.4 does not tell us anything about which of the extremely large number of identical lattice periods it may be in. The "orbit" of the electron may be very complicated; but whatever it looks like in one crystal cell, it must look like in any other one. On the *laboratory size scale*, the statement that the electron has a state corresponding to a particular energy does *not* tell us anything about its location in space.

Here again we are seeing the fact that electrons which are energetically in the valence band, for example, are not really to be thought of as stuck in a particular valence bond; nor are those in valence bonds all required to have anything like the same energy. The filled bond should perhaps more properly be thought of as a place where *two* electrons are sure to be—but we have no right to insist that it is always the *same* two! Again, we must not forget that, in principle, electrons are indistinguishable.

Thus the energy-band diagram of Fig. 2.4 applies to the entire piece of any laboratory-scale sample of the crystal to which it refers, and does not relate to special locations within that sample. By *laboratory scale*, we mean here a volume which contains a large number of unit periods of the crystal structure, such a unit period occupying a volume of about 10^{-23} cm^3. Accordingly, if we are to assign any directional significance to the horizontal axis in an energy-band diagram, any changes which might be considered to take place along such a direction would have to occur very slowly with respect to a lattice period. There has to be enough of a periodic piece, with no substantial variation through it, to justify its representation by bands at all; and this requires that a substantial number of lattice periods be included. In most cases where such space variations are shown on energy-band diagrams, the requisite approximations are reasonably well justified—especially because the picture is usually employed only for qualitative rather than quantitative purposes.

2.3 BAND STRUCTURE OF AN "EXTRINSIC" SEMICONDUCTOR

It is now necessary to consider how the addition of various impurity atoms to an otherwise perfect crystal structure may be expected to alter the energy-band diagram for a semiconductor. The matter is not entirely straightforward, because, for one thing, the substitution of donors or acceptors for normal atoms is *not* made in a regular periodic way. The impurities are usually distributed very sparsely throughout the crystal, and at random in the periodic structure.

Moreover, the new feature added by a donor, for example, is that an electron (the extra one) can be "bound" to it without being either in a valence bond or completely free to wander throughout the rest of the crystal. In other words, an electron can have a state, the energy of which is neither in the valence band nor in the conduction band. This state corresponds to the modified hydrogen-like orbit we calculated in Eqs. 1.5 to 1.12, and has an energy associated with it that differs by only a small amount from that of an electron in the conduction band. If such a state were to be shown in the band diagram, it would therefore be shown at about 0.01 ev below E_c for Ge and 0.04 ev below E_c for Si (see Table 1.1). However, note that for an electron to have this energy, it is *not* permitted to be located in space at a normal crystal lattice period, but must be located in the general neighborhood of one of the impurity atoms. A donor atom is, of course, locally neutral when the donor state is *occupied* by an electron. Therefore, an energy level representing the bound donor state implies more about space-location of an electron in that state than does any of the regular levels elsewhere in the band diagram. Accordingly, it is customary to show such states as short dashes, as indicated by the level E_d in Fig. 2.5*a* for a material with only donors added to it. The regular levels within the conduction and valence bands are shown as solid lines (to suggest the lack of localization associated with them, as discussed before).

The resulting picture in Fig. 2.5*a* should not be taken too literally, however, because the particular donor state we calculated in Chapter 1 was only the lowest orbit that the extra electron could have if it were bound to the donor. Clearly it could have many more excited states, just as a hydrogen atom can, and these levels

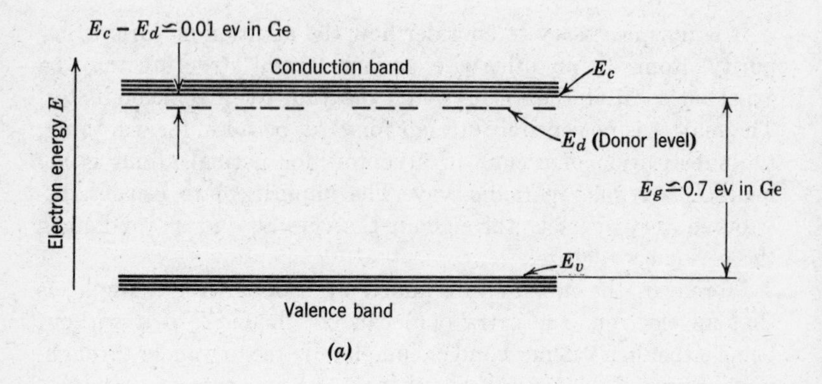

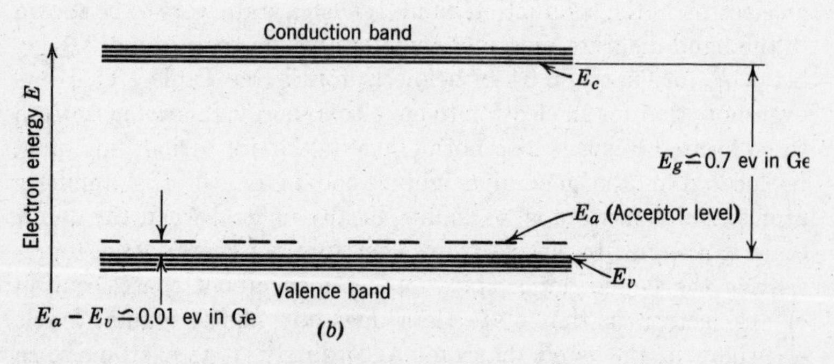

Fig. 2.5. Electronic energy bands for impurity semiconductors; (*a*) *n*-type, with donors only; (*b*) *p*-type, with acceptors only.

would appear between the lowest one and the conduction band. What has really been done is to remove from a perfect crystal one of its normal atoms, with all the electron states it contributes to the various bands, and substitute for it an atom with a somewhat different state structure *and* an extra electron. The result is a considerable disturbance of the local situation, as well as the introduction of an additional set of states for the extra electron. Fortunately, we shall find that at moderate temperatures circumstances are normally such that the extra electron is set free into the conduction band, so the details of the energy levels around the donor are not important. Even at low temperatures, where the "fine structure" we are discussing can actually be detected by special optical experiments, the lowest impurity state that we have calculated (see Eq. 1.12) still dominates the situation because it *is* the lowest state and is more likely to be occupied for that reason (P2.2).

The modification of band structure for inclusion of an acceptor impurity is shown in Fig. 2.5*b*. The lowest electronic state E_a added by such an atom is a little above the top of the valence band E_v. It is above this level because, as we have said, an electron added to the acceptor to complete its bond structure is only a little less tightly held from being free than it would be in a normal bond. The corresponding energy difference is again about 0.01 ev for Ge and 0.05 ev in Si (Table 1.1). The acceptor atom is locally neutral when the acceptor state is *not* occupied by an electron (i.e., when that state is *empty*).

Of course, any real sample of semiconductor is not perfect and will have a few of both donor and acceptor atoms in it (perhaps by accident, but often on purpose). The net result is easier to discuss, now that we have in mind the relative electron energies of the system as outlined in the band structure, than it was before when we considered it in Chapter 1. Inasmuch as an electron attached to an acceptor atom has a lower energy than one either in the conduction band or on a donor, the natural tendency is for the electrons available from either the conduction band or a donor to fill up the acceptor levels as much as possible. Certainly, at a temperature of absolute zero the conduction band would be empty, the valence band full, and the acceptor levels filled as far as possible by electrons made available by any donor atoms present. This situation is without doubt the lowest in system electronic energy. Therefore, if there are more acceptor atoms than donor atoms, the extra electrons available from all the donors will attach themselves to acceptors and *not* be available to enter the conduction band easily. Since the acceptors filled by this process are no longer able to accept further, they will not easily provide holes for the valence band. The result is that only the *difference* between the number of acceptors and the number of donors is available to provide holes easily for the valence band at normal temperatures.

Similarly, if the number of donors exceeds the number of acceptors, only the *excess* of donors over acceptors will have available their fifth electrons at zero temperature, and only these can easily supply free electrons to the conduction band at normal temperatures.

It is also true in principle that, at temperatures above zero, some electrons could in effect be excited from the acceptor level up to the donor level. But for the materials we shall be dealing with, the energy difference $E_d - E_a$ is so much larger than thermal energy kT, *and* the number of impurity atoms is so much smaller than the number of normal ones, that the excitations suggested are absolutely negligible compared to either the excess number of impurities or normal hole-electron pairs produced. Therefore, under the normal conditions of materials and impurity concentrations of interest here, it is the *excess* impurity that acts to produce carriers as if it were present alone.

It has been emphasized that the concentration of impurity atoms in the crystal lattice is usually small, and that the location of these atoms is random. Correspondingly, the electron having donor-level energy, E_d, for example, must be localized roughly at one of the donor atoms. Surely this is true if the donor atoms are far apart compared to the radius we have computed in Eq. 1.10 for the orbit of the fifth electron. If, however, a substantial number of the donors come closer together than this distance, we would expect to find them sharing each other's fifth electrons. When this happens, the extent to which the electron is bound to any one location also becomes less. The proximity of charged neighbors reduces the field binding electrons to any single atom. Therefore, it is not surprising to discover that the apparent "ionization energy" of the donors becomes smaller with increasing impurity concentration, in much the same way that the ionization energy of a normal atom is less when it is in the crystal than when it is isolated in space. Correspondingly, the donor energy level begins to split up into a noticeable group of levels, because of the coupling. A similar argument can be made for acceptors; it is interesting to do this in terms of the idea that an acceptor normally has a *hole* "bound" to it!

A rough estimate of the impurity concentrations at which mutual sharing of donor electrons should be observed, may be made as follows. Recall, for instance, that the orbit radius of a donor's fifth electron (or an acceptor's fifth hole) in Ge was found in Sec. 1.4.1 to be about 16A. For an average impurity spacing of about twice this radius, each impurity atom "occupies" a crystal volume of $(2 \times 16)^3 \times 10^{-24}$ cm^3 = 3.2×10^{-20} cm^3. The corresponding concentration of the impurities is the reciprocal of that volume, or about 3×10^{19} cm^{-3}. At about this level of "doping," Ge would certainly be expected to show some effects arising from overlapping impurity orbits. Actually, our calculations in Sec. 1.4.1 gave too large an ionization energy and too small an electron orbit. Because of this fact, and because of the irregularity of the spacing of impurity atoms in the crystal, some lowering of the effective ionization energy of the impurities may be expected to occur at much smaller *average* spacings than those implied by the concentrations described above. In fact, some such lowering is noticeable at concentrations as low as 10^{15} atoms/cm^3.

An interesting extreme form of the aforementioned splitting of the impurity levels sometimes occurs in material that has a large energy gap and large impurity-ionization energies. Such material normally is like an insulator; but at extreme doping levels the impurities may be close enough together on the average so an electron can just move from one to the other, as in a new "band," without entering the conduction band at all. This kind of "hopping" effect is referred to as "impurity-band" conduction. It is usually distinguished by low carrier mobility, *very* small Hall effect, and the persistence of substantial conductivity down to very low temperatures, as in a metal.

When the energy gap and the impurity ionization energies are not quite so large, however, so the material is *not* like an insulator in character, the value of the effective mass of the conduction electrons (for example) may be small enough to make their "size" in the lattice rather large (see Secs. 1.3.2 and 3.4.1). They then interfere *with each other* significantly, even at concentrations small enough so the orbits of carriers bound to adjacent impurities would *not* overlap each other. The conduction electrons then do not behave as *independent* gas-like particles any longer, and, as we shall examine more carefully later in Sec. 3.4.1, the quantum-mechanical effect of the Pauli exclusion principle must be felt. Such material is said to be "degenerate." It usually has moderate carrier mobility, moderate Hall effect, and substantial electrical conductivity, all of which tend to be rather insensitive to temperature over a wide range—again more nearly like the behavior of a metal than a normal semiconductor (see sample 58, Figs. 1.7, 1.11, and 1.13).

Of course, if the impurity ionization energy and the effective mass are small enough, not only will electrons in the conduction band interfere with each other, but electrons on donors will also interfere with each other and be "shared" enough to make the effective impurity ionization energy go to zero. The donor levels then split, and actually overlap states in the conduction band. The donor electrons are then free to wander in the conduction band—a process which they cannot distinguish from orbiting about a group of impurities. Such a material is also said to be degenerate, and it has the characteristics just described for degenerate materials. Indeed, it is not often easy to distinguish the two types of degeneracy described above, and in many materials both occur under similar conditions of doping. This is the situation for Ge at room temperature. The effective radius of a conduction electron in Ge at room temperature was found in Sec. 1.3.2 to be about 30A. So degeneracy from mutual interference of these carriers occurs at a concentration of about 10^{19} cm^{-3}, which is about the same as we found for overlapping of the orbits associated with the impurity states (P2.3).

There is no doubt that the properties of heavily doped semiconductors are more complicated than those of the lightly doped ones. Nevertheless, degenerate materials are of some practical interest, because of their nearly metallic conductivity coupled with the presence of an energy gap. The fact that the degeneracy may arise from either donors or acceptors, giving high conductivity by either free electrons or holes, has made possible such interesting devices as the Esaki tunnel diode.

2.4 ARRANGEMENT OF STATES IN THE BAND STRUCTURE

2.4.1 *Allowed Momenta*

Quantum-mechanically we should, in general, describe electron states by quantum numbers, whereas classically we would use momentum and position components. In the present problem, for states in the valence and conduction bands, the situation is actually intermediate. We have already assumed, without the necessary quantum-mechanical proof, the validity of a semiclassical description of the dynamical behavior of conduction-band electrons and valence-band holes, based on the use of the effective-mass idea. Thus a discussion of the states also based upon the validity of the same effective-mass representation of the carriers should be acceptable. It conserves the number of assumptions in a consistent way. Let us proceed on that basis.

What do we mean by an "almost free" classical-like particle in the crystal? We mean that such a particle is not bound to any particular atom—but may equally well be in the neighborhood of any of a large number of atoms in *equivalent* positions in the lattice. All we can really say with certainty about its location in space is that it is inside the volume of crystal we are considering. Let this volume have edges of length L_x, L_y, L_z, and hence volume $V = L_xL_yL_z$, as shown in Fig. 2.6*a*. Now on a *completely* classical basis, the fact that the electron is within a limited volume would not interfere at all with our ability to specify the momentum vector of the particle as precisely as we desire. In quantum mechanics, however, it is regarded as *conceptually impossible* to consider together the precise position and momentum of any single particle. Indeed, the situation is even more specific, in that the Heisenberg uncertainty principle relates quantitatively the lack of precision with which the position and momentum coordinates may be specified together, as follows.

$$\begin{aligned} \Delta x \Delta p_x &\geq h \\ \Delta y \Delta p_y &\geq h \\ \Delta z \Delta p_z &\geq h \end{aligned} \tag{2.1}$$

The result is that a precise statement about the location x, which makes $\Delta x = 0$, requires a total doubt about the momentum component p_x in the sense that $\Delta p_x = \infty$.

Thus our statement of certainty that the particles in a material are somewhere inside the volume V with dimensions given above, and of complete ignorance of further detail about their locations,

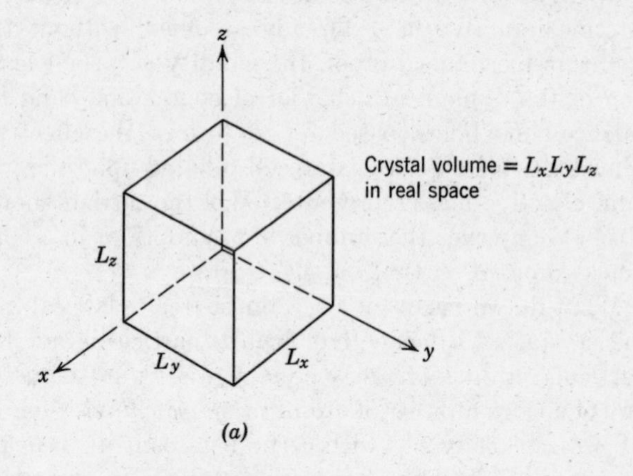

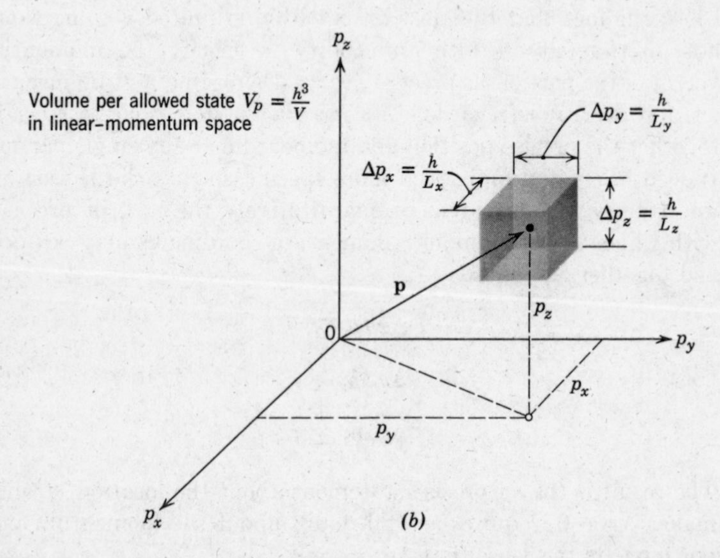

Fig. 2.6. To determine the number of dynamical states per unit volume of crystal we need (*a*) the crystal volume V in real space; (*b*) the corresponding volume V_p per allowed state in linear momentum space.

implies that we cannot possibly conceive of their momentum components within a precision range

$$\Delta p_x = \frac{h}{L_x}$$

$$\Delta p_y = \frac{h}{L_y} \tag{2.2}$$

$$\Delta p_z = \frac{h}{L_z}$$

In terms of the vector momentum **p**, we can understand the implications of this result if we will construct a diagram along whose axes are plotted the components of momentum. Such a diagram defines a "momentum space," in which it is convenient to show the momentum vectors of one or more particles. A sample for our present use appears in Fig. 2.6*b*. The results in Eq. 2.2 mean that it is *in principle* impossible to state the value of the momentum vector **p** any more precisely than that it ends somewhere inside the little momentum-space box of volume

$$V_p = \frac{h^3}{L_xL_yL_z} = \frac{h^3}{V} \tag{2.3}$$

Accordingly, the only valid quantum description of the momentum vector **p** is that its end lies in a little box which is itself located by stating that it is the n_xth box along the x-axis, the n_yth along y, and the n_zth along z! That is, the entire momentum space itself must be thought of as being built out of indivisible bricks, each of volume V_p, and it must be regarded as nonsense in principle even to speak about *points* in momentum space on any finer scale. It is in this way that we come to describe the linear momentum of a particle in quantum mechanics by using only three whole "quantum" numbers n_x, n_y, and n_z.

Actually, as we have said before, it takes four quantum numbers to specify the state of an electron, its two possible spin values being the remaining choice. Because of the Pauli exclusion principle, however, no two particles can have the same four quantum numbers. In the present case this means that *not more than two electrons can have a linear momentum within any one box* V_p, *and, to get two into the same box, they must have oppositely directed spins.* There are just *two available electron states* for each volume $V_p = h^3/V$ in linear-momentum space.

2.4.2 *Relation between Energy and Momentum*

Next we need to know the electron energies corresponding to the various allowed states. That is, we must know the relation between the momentum and the energy of the electrons when they are in the conduction or valence bands.

Simple as it may seem, this question would not be straightforward for particles which really had to be regarded completely from the point of view of quantum mechanics. Because of the uncertainty principle (or the wave nature of matter, if you prefer to think of it that way), the usual connection between momentum and kinetic energy that involves directly the particle velocity, and indirectly the idea of precise position versus time, is not valid any longer. The whole notion of describing the position of a particle as a function of time, then finding its velocity as a function of time, and finally making its momentum just $m\mathbf{v}$, would lead us straight into giving position and momentum together precisely! In quantum mechanics, therefore, the connection between energy and momentum of a particle in the general case is not as simple as it is classically.

In Sec. 1.3.3 we have, however, tried to point out that a true quantum-mechanical solution for the response of an electron *in the conduction band* to a weak externally applied force leads to the novel conclusion that it behaves approximately like a classical particle with an *effective* mass m_e^*. Inasmuch as an electron is not in the conduction band at all unless its total energy E exceeds E_c (Fig. 2.5), the sense in which this equivalent classical representation applies is

$$\left.\begin{aligned} &(a) \qquad E = E_c + \frac{|\mathbf{p}|^2}{2m_e^*} \\ &(b) \qquad \mathbf{p} = m_e^*\mathbf{v} \end{aligned}\right\} \begin{array}{l} \text{electrons with} \\ E \geq E_c \\ \text{(conduction band)} \end{array} \tag{2.4}$$

Correspondingly, recall that a hole behaves like a classical particle too, but with effective mass m_h^*. That is, after it is once produced, its energy of motion (kinetic energy) is

$$\left.\begin{aligned} (a) \qquad E_{\text{hole}} &= \frac{|\mathbf{p}|^2}{2m_h^*} \\ (b) \qquad \mathbf{p} &= m_h^*\mathbf{v} \end{aligned}\right\} \text{holes in valence bond} \tag{2.5}$$

Now, if an electron is just barely raised in energy from the level E_v at the top of the valence band to the level E_c at the bottom of the conduction band, a free electron e_1 results at rest in the conduction band, while a hole h_1 remains at rest in the valence band, as shown in Fig. 2.7. The system has been given an increment of energy $\Delta E = E_g$.

If, next, the energy of the free electron is increased by giving it some velocity, leaving the hole undisturbed, Eq. 2.4 shows that the system energy has changed by $E_g + |\mathbf{p}|^2/2m_e^* = \Delta E$. The situation becomes that shown in Fig. 2.7 by e_2 and h_2, where a hole-electron pair is produced with the hole at rest and the free electron moving with momentum $\mathbf{p}$.

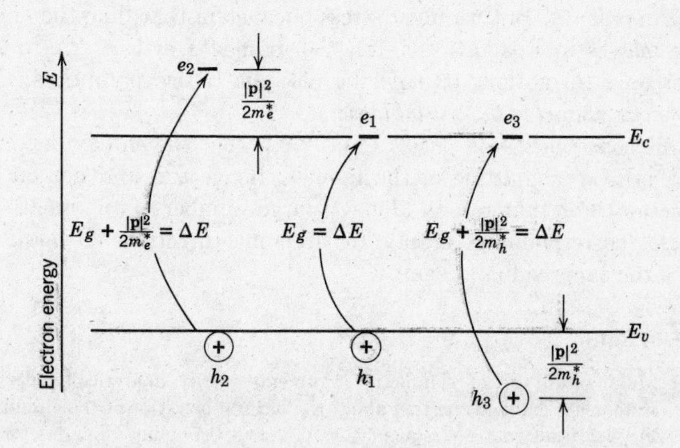

Fig. 2.7. Energy-momentum relations in the band structure.

Suppose, however, that we started with the pair at rest, and then gave the hole some momentum $\mathbf{p}$, leaving the free electron at rest. Equation 2.5 shows that we should obtain a total increase of system energy equal to $\Delta E = E_g + |\mathbf{p}|^2/2m_h^*$ over the case without the pair at all. But inasmuch as all such changes in total energy must be accounted for by *electron* behavior in either the valence or conduction bands, an electron must actually have been raised from an energy *below* E_v by the amount $|\mathbf{p}|^2/2m_h^*$ just to the bottom of the conduction band, to account for the "kinetic energy" of the hole (h_3 and e_3 in Fig. 2.7). The clear implication, then, is that there must have been an electron down in the valence band occupying a *state* that had an energy *below* E_v by the amount $|\mathbf{p}|^2/2m_h^*$, as indicated below.

$$\left.\begin{aligned} &(a) \qquad E = E_v - \frac{|\mathbf{p}|^2}{2m_h^*} \\ &(b) \qquad \mathbf{p} = m_h^*\mathbf{v} \end{aligned}\right\} \begin{array}{l} \text{electrons with} \\ E \leq E_v \\ \text{(valence band)} \end{array} \qquad (2.6)$$

Equations 2.4 and 2.6 tell us, within the approximate viewpoint we have adopted for hole and free electron behavior, how the energy of *electrons* in the conduction and valence bands depends on their momenta. But we must stress once again that, like the effective masses in Eqs. 1.3 and 1.4, the momenta in Eqs. 2.4 to 2.6 refer only to motions *through* the periodic lattice potentials; i.e., only *with respect to the crystal lattice.*

One consequence of Eqs. 2.4 and 2.6 is that the energy depends only on the magnitude of the momentum vector, and not on its direction. For that reason alone, a large number of (momentum) states, corresponding simply to different directions of motion, have the same value of energy.

PROBLEMS

P2.1 Making full use of symmetry arguments, verify analytically all the statements made in the text about the natural behavior of the circuit of Fig. 2.2. Construct a "frequency level" versus "coupling" diagram for it.

P2.2 Calculate the approximate binding energy and orbital radius of an electron on a donor atom in the first excited state. Can this state be observed?

P2.3 In semiconductors with large energy gaps, the relative dielectric constant tends to be small, whereas in those with small energy gaps it tends to be large.

(*a*) In terms of the valence *bond* model of the material, does this seem reasonable? Why?

Also, the effective masses tend to be large in large-gap semiconductors, and small in those with small gaps.

(*b*) In terms of the strength of the crystalline periodic forces, does this seem reasonable? Why?

(*c*) Express the radius r_{oi} of the orbit of an electron bound to a donor in terms of the relative dielectric constant $K \equiv \epsilon/\epsilon_o$ and the normalized effective mass $\nu \equiv m_e^*/m$. Reduce everything else to a univeral numerical constant.

(*d*) Express the "effective radius" r_{oe} of a conduction-band electron as a function of $\nu \equiv m_e^*/m$ and the normalized temperature $\eta \equiv T/300$. Reduce everything else to a universal numerical constant.

(*e*) Express the donor ionization energy E_o as a function only of ν and K. Locate Si and Ge on plots of E_o versus ν for two values of K.

(*f*) Plot r_{oi} and r_{oe} versus ν for $K = 16$ and $\eta = 1$. Show where Ge is on these plots.

(*g*) Repeat part (*f*) for Si.

(*h*) Consider three *n*-type semiconductors (with donors only): Ge, one with a considerably larger energy gap, and one with a considerably smaller energy gap. In view of the correlations pointed out in parts (*a*) and (*b*), and your plots in parts (*c*) through (*g*), discuss for each material the kind of "degeneracy" that may occur at $T = 300°K$, and find the impurity concentration at which you expect it to occur in each case.

(*i*) What effect does changing the temperature have on your answers to (*h*) above?

P2.4 (*a*) What is the longest wavelength of light, a photon of which can produce a hole-electron pair at rest in Ge? In Si?

(*b*) Consider holes and free electrons in Ge at room temperature. If pairs were to recombine by each giving up their entire energy as a single photon of light, what is the longest wavelength you would expect to find? What would be, approximately, the *average* photon energy observed? (See also P3.2, and the film referred to on p. 59.)

3

The Equilibrium Distribution of Electrons in the Bands

3.1 DISTRIBUTION FUNCTIONS

Having described the general form of the energy-band structure of the available states for electrons in a crystal, we must next investigate how the electrons behave under various conditions, such as in thermal equilibrium at a temperature T, or in response to disturbing influences like an electric field, a magnetic field, or incident light. In this chapter we devote ourselves primarily to the condition of thermal equilibrium, reserving for Chapter 4 the consideration of nonequilibrium phenomena.

At first thought, determination of the properties of a system of interacting particles would seem to require a solution in detail for the dynamics of each particle. Even if it were really desirable to make such a solution, however, the fact that there may be 10^{23} particles per cm^3 would make the job incredibly complicated to carry out. Moreover, we really would not know what to do with the "answer" after it was written down, because it would cover a universe of paper.

Fortunately, we usually only care to make measurements of currents, charges, voltages, pressures, and other similar quantities that involve either a large number of particles or net action over a relatively long time. Thus our results depend only on the average behavior of a single particle rather than its detailed actions at any moment. Average individual behavior, or the behavior of large groups of particles, then, assumes great importance in our observations (measurements) of the physical world. The underlying detailed variations only intrude themselves (unpleasantly enough, however) in the form of various kinds of "fluctuation noise."

The points just raised are rather practical reasons why we might be tempted to adopt a statistical view of systems comprised of many bodies; but there is another good reason why we are well-advised to give up a detailed description of these complicated systems. It is connected with the fact that we always consider the properties of a *localized* system—a sample of material—to which we may apply certain local forces under our control, and then wish to observe the response. The difficulty is that there are other, nonlocal, forces involved which we do not know how to describe in detail. These are the ones comprised of such things as the banging of air molecules against our sample, or the banging of light photons from the sun on it, etc. There is no way to select a segment out of the universe, so that we have access to it ourselves, without having the rest of the universe exert some sort of influence on it in a manner outside our detailed knowledge and control. This is the main reason why we need the ideas of heat and temperature, and why there is connected with them the notion of randomness which we always use to describe things whose details we do not know. Thus at the very outset, we face random forces acting on our materials to establish their "thermal environment," even before we apply any additional forces to them. Accordingly, a statistical description of the random properties of the system becomes not only entirely appropriate but also virtually necessary.

Because it is the simplest, let us consider first the situation that prevails when the external random environment acts alone, and has been doing so for a very long time. The condition we have in mind is called "thermal equilibrium," and is characterized (among other things) by a temperature T degrees Kelvin. Without trying to be complete or quantitative, we can point out a number of qualitative features associated with the state of thermal equilibrium: it represents a "settled" condition, with no detectable steady trends as a function of time; if several constituents are present together, the situation has become as thoroughly "mixed" and independent of past history as possible; the condition of the system under consideration has become as "diffuse" or "disorganized" as it ever will, without the imposition of new forces. All of these ideas relate to the fact that any initial organized arrangement of the system has had time to "relax" away as far as possible, while the system has been bombarded steadily by its environment at temperature T.

Whereas we are not able to describe in detail what happens to any particular particle as a function of time, we have pointed out that, by considering suitable average behavior, we may determine with good accuracy what large numbers of particles will do. Let us be more specific about the meaning of "average behavior" in connection with dynamical problems.

First, the description of particle dynamics in classical systems requires the specification of momentum and position, whereas in quantum-mechanical systems it requires the specification of quantum state.

Next, for the statistical rather than the detailed treatment, we select a certain dynamical behavior instead of a certain particle. For instance, we might select the behavior described by specifying a momentum of magnitude less than p, and a position inside radius r from some origin. Then we may ask how many particles *in a given system* will, *on the average*, have this specified behavior.

The "average" involved in the question posed above may be looked upon fundamentally in two different ways, which in the vast majority of physical problems may be treated as being completely equivalent.

One view supposes that we observe the given system for a very long time and, in particular, observe the temporal history of the number of particles whose dynamical behavior is of the specified type. The *time average* of this number function is then the desired one.

Alternatively, we may consider a very large number of systems whose measured properties are identical to the one given. Then, at an arbitrary single moment we determine how many of the particles in each system have the specified dynamical behavior. The arithmetic or *ensemble average* of these numbers is again the desired one.

We are led then by the foregoing considerations to the idea of a *distribution function.* For a given system containing N particles, at equilibrium temperature T, such a function gives us the average number of particles that will have each and every specified dynamical behavior.

In respect to these distribution functions, there are two differences between classical and quantum-mechanical systems that need attention right away.

In classical systems, the momentum $\mathbf{p}$ and the position $\mathbf{r}$ are continuous variables, and the dynamical state is specified by giving the values of both at any time. Presumably, then, the distribution function might be imagined to give the average number of particles that will have *exactly* any selected momentum $\mathbf{p}$ and position $\mathbf{r}$. But such a function is badly misbehaved (not continuous). Consider momentum alone, for instance. If the distribution suggested above were reasonable, in the sense that it had nonzero positive values and were continuous over a small momentum range $\Delta\mathbf{p}$, the total number of particles having momenta in this range would be *infinite*, corresponding to the infinite number of precise values of $\mathbf{p}$ in this range! It clearly does not make sense here to try to use functions restricted by this sort of difficulty.

The kind of distribution that does make sense under the above circumstances is one which gives the average number of particles, known to be in a given region of space, for example, that have a momentum within some finite limits, as for instance with an x-component between p_{x1} and p_{x2}. We could also add the idea of a distribution in space by giving a range of positions between x_1 and x_2, etc. In any case, the total number of particles having all possible vector momentum values and all possible vector positions in the system would have to be equal to N, the total number of particles in the system considered. So the distribution function must be dependent on N, as well as on T.

Actually, whereas the distribution function we have just described is the most straightforward one, it is often more convenient to deal with a differential distribution that amounts to the derivative of the one defined over finite intervals. The concept in question is called the *distribution-density function* $f(p_x, p_y, p_z; x, y, z)$. This function is defined so that the number of particles in the differentially small dynamical range between p_x and $p_x + dp_x$, p_y and $p_y + dp_y$, p_z and $p_z + dp_z$; x and $x + dx$, y and $y + dy$, and z and $z + dz$ is just $f dp_x\, dp_y\, dp_z\, dx\, dy\, dz$. Accordingly, this number vanishes as we try to specify $\mathbf{p}$ and $\mathbf{r}$ exactly by driving $d\mathbf{p}$ and $d\mathbf{r}$ to zero. However, we know that

$$\int_{-\infty}^{\infty} \cdots \int_{-\infty}^{\infty} f(p_x, p_y, p_z; x, y, z)\, dp_x\, dp_y\, dp_z\, dx\, dy\, dz = N \qquad (3.1)$$

because, again, the number of particles located *somewhere* in the system, and doing *something*, has to be all those included in the system.

In many cases, the material being treated is homogeneous in composition, so the distribution density does not depend on position. Under these circumstances, the space coordinates may be omitted from consideration and only the momentum components are needed. The result $f(p_x, p_y, p_z)$ then describes momentum behavior for any desired volume of space (for example, per unit volume), and no integration over space coordinates is needed to insure that the distribution includes the necessary N particles *per volume considered.*

The situation in quantum mechanics differs from that described above in *two* important ways. First, each dynamical state of a particle is specified by only 4 quantum numbers, including spin, rather than by 6 components (3 for position and 3 for momentum). Second, the description of dynamical state is *discrete* rather than continuous. Thus we ask quite legitimately for the average number of particles that have a particular dynamical state identified by four discrete quantum numbers n_1, n_2, n_3, n_4. If $f(n_1, n_2, n_3, n_4)$ is the function which describes that average number, then

$$\sum_{\substack{\text{all}\\ n_4}} \sum_{\substack{\text{all}\\ n_3}} \sum_{\substack{\text{all}\\ n_2}} \sum_{\substack{\text{all}\\ n_1}} f(n_1, n_2, n_3, n_4) = N \qquad (3.2)$$

because all the particles must be distributed somehow among all the allowed states of the system under consideration. For quantum-mechanical problems, f is therefore permitted to be a *distribution function* directly.

3.2 BOLTZMANN DISTRIBUTION FOR NONINTERACTING ENTITIES

3.2.0 *Introduction*

In either the classical or quantum-mechanical cases, the form of the distribution function (or density) f depends critically on the extent to which the particles, or other entities being described, influence each other. Of course, the temperature also enters as a dominant feature, inasmuch as this parameter describes the random features of the environment external to the systems under consideration. It is not surprising that the very simplest situation is the one in which the "particles" in question do not interact with *each other* at all, but are all in contact with the external environmental temperature T.

Then the distribution must depend only on the dynamical variables of each separate particle, the number of particles, and, of course, on the temperature. The question is: What is the form of this distribution, so characteristic of the temperature?

Instead of trying to derive from very general postulates the answer to the question just posed, let us investigate one familiar example of a system of noninteracting particles and draw from it the answer we need. Specifically, we shall use our particular knowledge of ideal gases to *make plausible* (*not* to derive) the correct distribution function for *any* noninteracting entities which are in thermal equilibrium at absolute temperature T.

3.2.1 *The Ideal Gas under Gravity*

For the purpose described above, consider a cylinder of ideal gas, standing under the influence of gravity, in an environment at temperature T (Fig. 3.1). The molecules of such a gas do not interact with each other (they need not even collide with each other), but they *must* collide with the walls of the system at temperature T. Thus let us think of each molecule (which may in the most general circumstance be composed of a number of atoms) as being an example of *any* entity, however complex, which is in thermal equilibrium among other similar ones with which it does not interact. We take the vertical position z as the dynamical variable in terms of which the distribution function will apply. In this case of the ideal gas, we have a classical system; so we shall actually get the answer for the classical case. The result we need is the fraction of molecules to be found in the height interval between z and $z + dz$. How to modify the results for quantum-mechanical situations will become clear as we proceed.

Let $n(z)$ be the number of molecules per unit volume at height z. We wish to determine $n(z)$. Also let m be the mass of each molecule

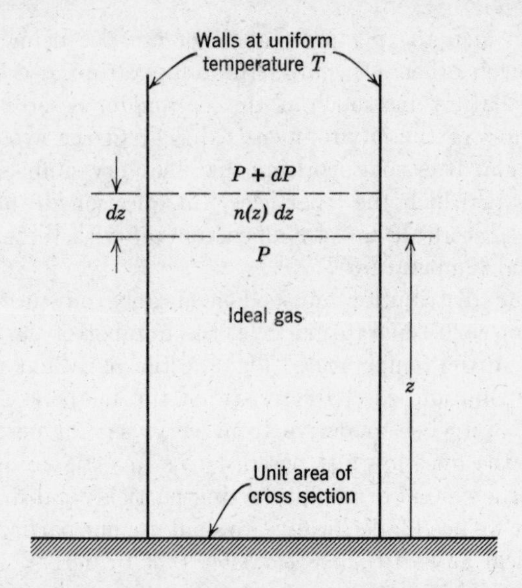

Fig. 3.1. Ideal gas under the action of gravity in an isothermal cylinder.

(which we treat as a simple point-particle for this example), g be the acceleration of gravity, and ρ be the density of the gas. Then consideration of the pressures on the top and bottom of the slab of gas of thickness dz shown in Fig. 3.1 yields

$$dP = -\rho g\, dz = -nmg\, dz \tag{3.3}$$

We must find another relation between $P(z)$ and $n(z)$ to solve for $n(z)$. Such a relation comes from the law of ideal gases,

$$PV_M = RT \tag{3.4}$$

where V_M is the volume of one mole of the gas, and R is the gas constant per mole† ($R = 8.31$ joules per mole per °K). The volume V_M is related to n through Avogadro's number ($N_o = 6.025 \times 10^{26}$ molecules per mole), because the mass of a mole of gas is $N_o m$ and we have

$$V_M = \frac{N_o m}{\rho} = \frac{N_o}{n} \tag{3.5}$$

† A mole in the mks system is that amount of a given material for which the mass in *kilograms* is equal to the molecular weight of the constituent molecules.

Thus, from Eqs. 3.4 and 3.5, we find for the second relation we needed between P and n:

$$P = n\left(\frac{R}{N_o}\right)T = nkT \tag{3.6}$$

in which we have written Boltzmann's constant k in place of the ratio R/N_o to which it is equal by definition.

To eliminate P between Eqs. 3.3 and 3.6, we first differentiate Eq. 3.6 with respect to z, under the condition that T is independent of z:

$$\frac{dP}{dz} = kT\frac{dn}{dz} \tag{3.7}$$

and then substitute into Eq. 3.3 to eliminate dP/dz. These steps give us the necessary differential equation for $n(z)$ alone.

$$-\frac{dn}{n} = \frac{mg}{kT}\,dz \tag{3.8}$$

Integration of Eq. 3.8 is easy:

$$n = Ae^{-mgz/kT} \tag{3.9}$$

where A is a constant independent of z (but possibly dependent on T and N).

Notice now that the term mgz in the exponent of Eq. 3.9 is the energy (purely potential energy in this example) associated with the coordinate interval z. So the number of molecules per unit volume at the height corresponding to energy $E(z) = mgz$ is just

$$n(z) = Ae^{-E(z)/kT} \tag{3.10}$$

Without loss of generality, $E = 0$ has been chosen arbitrarily at $z = 0$.

Because z is a continuous variable, the vertical distribution of molecules must be described by a distribution-density function. This function is precisely $n(z)$ because $n(z)\,dz$ is the number of molecules in the height range dz at z (the cross section of the cylinder in Fig. 3.1 being of unit area). Then Eq. 3.10 is the distribution-density function for coordinate z.

Now it is important to observe from Eq. 3.10 that the ratio of the numbers of particles at two different heights is a simple negative exponential of the energy difference between these heights, measured in units of kT. But to learn more, we must eliminate the arbitrary constant A. This we do by requiring that $n(z)$ sum (or integrate) over all z to the value N, as described by Eq. 3.1. Thus, from Eq. 3.10 and the definition of $E(z)$ for this problem, we find

$$\int_{z=0}^{\infty} n(z)\,dz = A\int_{0}^{\infty} e^{-mgz/kT}\,dz = A\left(\frac{kT}{mg}\right) = N \qquad (3.11)$$

Placing the solution for A from Eq. 3.11 into Eq. 3.10, we find

$$n(z) = \left(\frac{Nmg}{kT}\right)e^{-mgz/kT} = \left(\frac{Nmg}{kT}\right)e^{-E(z)/kT} \qquad (3.12)$$

The form of Eq. 3.12, which adds up properly to the total number of particles in the system, is said to be the "normalized" form of Eq. 3.10.

If we now examine the normalized distribution-density function in Eq. 3.12, we see that it has the form

$$f(z)\,dz = A(N, T)e^{-E(z)/kT}\,dz \qquad (3.13)$$

in which A is a parameter that indeed depends on both N and T, but *not* on the coordinate z with which the energy varies.

3.2.2 *Generalization*

We shall now generalize Eq. 3.13 by assuming that it applies separately to each and every dynamical variable (momentum and position) of the noninteracting entity being considered. As we have been suggesting, these "entities" need not be as simple as point-particles. For instance, we might have a molecule with M atoms in it. If treated classically, these M atoms are thought of as masses held together (interacting) by "springs." Each *molecule* then has $6M$ dynamical variables needed to specify its whole dynamical state: 3 position coordinates and 3 momentum components for *each* of M atoms. So if the whole molecules don't interact with each other (i.e., if they behave as ideal-gas molecules), we shall take it as a *postulate* that in thermal equilibrium at temperature T, the number of such molecules in a dynamical state in the neighborhood of position coordinates $x_1 \ldots x_{3M}$ and momentum components $p_1 \ldots p_{3M}$ is just

$$f\, dx_1 \ldots dx_{3M}\, dp_1 \ldots dp_{3M} =$$

$$A(N, T) \exp\left[-\frac{E(x_1 \ldots x_{3M}; p_1 \ldots p_{3M})}{kT}\right] \qquad (3.14)$$

$$dx_1 \ldots dx_{3M}\, dp_1 \ldots dp_{3M}$$

where E is the energy that the molecule (entity) has when it is in the specified dynamical state $(x_1 \ldots x_{3M}; p_1 \ldots p_{3M})$, and N is the total number of molecules in the system being considered. Notice here that, with so many variables ($6M$), the labeling scheme has been modified for convenience so that x_1, x_2, x_3 are, respectively, the x, y, z coordinates of one atom in the molecule, x_4, x_5, x_6 the coordinates of a second one, and so on through the Mth atom. Similarly, p_1, p_2, p_3 represent linear-momentum components p_x, p_y, p_z, respectively, of the first atom, and so forth.

Note that it would *not* be appropriate to apply Eq. 3.14 directly to each *atom* in the molecule, because these atoms *do interact with each other* through springs. Equation 3.14 is taken to apply *only* to *noninteracting entities*, for the classical rather than the quantum-mechanical case, in which each such noninteracting entity has a dynamical state described by $3M$ independent position variables and $3M$ independent momentum components. (For motions constrained to one or two dimensions, of course, only one or two position coordinates and one or two momentum components are needed for each single particle involved.)

Thus the essential feature of the temperature in connection with the equilibrium state of noninteracting entities is seen to be the fact that in a given system of N entities *the occupancy of each dynamical state is weighted exponentially according to the energy of that state in kT-units*. As a corollary, *all dynamical states with the same energy have the same occupancy*. This is a reasonable, though not obvious, property for states whose details we agree never to observe too closely.

The foregoing features of the statistical behavior of any noninteracting entity in thermal equilibrium at temperature T can, at the very foundation, be taken to define precisely what "temperature" means. There are other ways to state the relation between the macroscopic idea of temperature and the statistical behavior of a system of particles; but in every case a pure postulate connecting the mechanics of the particles with the nonmechanical idea of temperature must be made, and then checked against experiments. We have chosen here to make that postulate by simply generalizing Eq. 3.13, without any proof, to the form of Eq. 3.14.

3.2.3 *An Example*

As a very simple example of the result in Eq. 3.14, take the entity as a monatomic molecule of an ideal gas, treating each molecule as a point particle of mass m, as we did in our first example. But this time let the gas vessel be short enough to permit neglecting the variation of potential energy with height. Then the only relevant question is the number of molecules that have momentum in the neighborhood dp_x, dp_y, dp_z about the values p_x, p_y, p_z. Because in this case

$$E = \frac{p_x^2 + p_y^2 + p_z^2}{2m} \tag{3.15}$$

and the position coordinates do not matter if the gas is homogeneous, we find from Eqs. 3.14 and 3.15:

$$f(p_x, p_y, p_z)\, dp_x\, dp_y\, dp_z = A(N, T) \exp\left(- \frac{p_x^2 + p_y^2 + p_z^2}{2mkT} \right) dp_x\, dp_y\, dp_z \tag{3.16}$$

To evaluate $A(N, T)$, we must insist that the distribution be normalized in the manner of Eq. 3.1. For this calculation we need to know from a table of integrals that

$$\int_0^\infty y^n e^{-y^2}\, dy = \frac{1}{2}\left(\frac{n-1}{2}\right)!; \qquad \text{integer } n > -1 \tag{3.17}$$

in which

$$\left(-\frac{1}{2}\right)! = \sqrt{\pi}; \qquad m! = m(m-1)! \tag{3.18}$$

We must carry out the integration of Eq. 3.16, according to the requirement of Eq. 3.1, by considering all three momentum components from $-\infty$ to $+\infty$. The required integrals appear in

Eqs. 3.17 and 3.18 if we take $n = 0$, and give us for the result of the three integrations

$$A(N, T) = N/(2\pi mkT)^{3/2} \tag{3.19}$$

Accordingly,

$$f(p_x, p_y, p_z)\, dp_x\, dp_y\, dp_z = \frac{N}{(2\pi mkT)^{3/2}} \exp\left(-\frac{p_x^2 + p_y^2 + p_z^2}{2mkT}\right) dp_x\, dp_y\, dp_z \tag{3.20}$$

which is the classical Maxwell-Boltzmann distribution of molecular momenta (or velocities, since $\mathbf{p} = m\mathbf{v}$ in such cases). (An example of particles with both potential and kinetic energies to be considered appears in P3.1*a* through *d*).

3.2.4 *Quantum-Mechanical Distribution*†

To modify our foregoing postulate to apply to a quantum-mechanical ase, we need only to recall that in quantum problems both the dynamical states and their corresponding energies are discrete. Thus, as we said before, it makes sense to ask for the *number* $f(j)$ (*not* the number per unit range) of the entities that have a state designated by j. Actually, we have pointed out that four quantum numbers $n_1 \ldots n_4$ are needed to define the "state" of a single particle, and for an entity composed of M such particles, $4M$ quantum numbers are needed. However, we often simplify notation by letting each value of a single number j stand for one distinct *combination* of 4 or $4M$ quantum numbers. Hence, in simplified notation, the quantum distribution function for noninteracting entities with states described by j is just

$$f(j) = A(N, T)e^{-E_j/kT} \tag{3.21}$$

where E_j is the total energy of the dynamical state described by index j. Note carefully that the *form* of Eq. 3.21 is quite *independent* of the arrangement of states available for the entities in the quantum-mechanical problem being treated, but the nature and arrangement of these states *will* have to show up when we try to evaluate $A(N, T)$ by summing $f(j)$ over all j as required by Eq. 3.2 (P3.1*e*). Evidently Eq. 3.21 is the quantum-mechanical form of Eq. 3.14.

† Sections like this, set in smaller type, may be omitted in an introductory study, without loss of continuity.

3.3 TRANSITION PROBABILITIES

By applying the principle of detailed balancing to our previous considerations, we can learn something from the case of noninteracting particles that actually has general validity for interacting ones as well. We refer to an important relationship between "transition probabilities," which we shall now discuss.

Consider any system of noninteracting particles which has discrete quantum states (P3.1*e* takes up a specific example, but here we can simply imagine a fictitious one like a "quantized" ideal gas). This means that in the absence of any disturbances like thermal agitation, the particles would occupy the various states—i.e., traverse the various allowed orbits—in a fixed, unchanging way. They do not interact with each other.

But if the system is put in contact with a "reservoir" at temperature T, the thermal activity of this environment knocks particles from one state to another in a random manner. In other words, the random banging of the environment changes the momentum and energy of the particles in a random way. Focus attention, then, on two of the particle dynamic states identified by indices i and j, respectively. Let E_i and E_j be the total energies of a particle in these states. *At thermal equilibrium, the average number of particles in a unit volume of the system that make transitions from state i to state j per unit time must be equal to that in the reverse direction, from state j to state i.* This requirement follows from the principle of detailed balancing for these two states.

To express the equality of the two rates above for a unit volume of the system containing many particles, we must use the probability W_{ij} that in a unit time an individual electron already in state i will go to the given state j. We shall not need to know W_{ij} in detail. It is sufficient to realize that it depends upon the two specific states involved, and also upon the temperature. The temperature enters because the transition will either absorb energy and momentum from, or emit them to, the environment, and the frequency of the exchange for a given pair of states clearly will depend on the thermal activity of the surroundings.

Now considering, say, a unit volume of our system, which contains N particles, we recognize that for a particle to make a transition from state i to state j it must first be *in* state i, and then make the transition.

But we already know that the average number of particles in state i for this system is $A(N, T)e^{-E_i/kT}$.

So, *per unit time of observation*, the average number of particles that make the transition from state i to state j is just

$$A(N, T)e^{-E_i/kT}\, W_{ij}$$

Similarly, the average number of transitions that take place in unit time the other way, from state j to state i, is just

$$A(N, T)e^{-E_j/kT}\, W_{ji}$$

the first factor being the average number of particles in state j, and the second factor the probability per unit time of a transition from j to i.

For the average number of transitions per unit time from j to i to balance those from i to j, the two transition rates just calculated must be equal. Thus

$$A(N, T)e^{-E_i/kT}W_{ij} = A(N, T)e^{-E_j/kT}W_{ji} \tag{3.22}$$

or

$$\frac{W_{ji}}{W_{ij}} = e^{(E_j - E_i)/kT} \tag{3.23}$$

The ratio of the equilibrium transition probabilities per unit time in both directions between two given states is exponential in the energy difference between the states, the larger probability applying to the direction of decreasing energy.

This conclusion is really quite important, because it relates the quantities W_{ji} and W_{ij} which, by their very definition, depend *only* on the dynamics of the given states i and j, and on the temperature. Specifically, these transition probabilities do not depend on interparticle interaction at all. The effect of such interaction may well be to cause particles to have a different set of permitted dynamical states than they would have had without any interaction; or it may limit the number of particles that may have any particular state. But once the set of allowed states for the particles has been defined, and we place a particle in one of them, then the process by which this particle goes into another available state, specified in advance, depends only upon the disturbing influences of the environment—i.e., the temperature—in addition to the two particular states involved, of course. Accordingly, Eq. 3.23 is actually more general than the noninteracting system we used as a vehicle to derive it! In fact, *the result in Eq. 3.23 applies to the particles of every system, whether or not the particles interact with each other*, provided only that it makes sense to talk at all about states for individual particles, and about transitions between such states produced by the random thermal banging of the environment on the system. Violation of this last restriction, however, only occurs in cases quite beyond the scope of our whole discussion here; so we may proceed to build upon Eq. 3.23 even for a discussion of particles like electrons in a solid, which obey the Pauli exclusion principle.

3.4 FERMI-DIRAC DISTRIBUTION FUNCTION FOR ENTITIES SUBJECT TO THE EXCLUSION PRINCIPLE

As we have said before, the form of the distribution function (or density) must depend greatly on the kind of interaction taking place between the entities involved. Because of the Pauli exclusion principle, the electrons in a solid are interacting with each other. They must avoid each other if any two find themselves moving in nearly an identical manner. Thus individual electrons in a solid cannot be taken as entities to which the Boltzmann distribution applies directly. The nature of the interaction in this case is, however, quite simple to state quantum-mechanically: no two electrons can occupy the same quantum state, if by "state" we include the three orbital quantum numbers and the spin.

Because of this simplicity, we can quite easily find out what the proper equilibrium distribution function is for electrons in the solid by appealing to the principle of detailed balancing, and our general result (Eq. 3.23) relating to transition probabilities. To do this, consider again two electronic states i and j, with energies E_i and E_j, respectively. The individual transition probabilities per unit time W_{ij} and W_{ji} are of the same nature as we have discussed before; but now, because of the exclusion principle, to make a transition from state i to state j we must not only have an electron in state i but we must specifically *not* have one in state j. Therefore, our W_{ij} is a probability of transition per unit time, *conditional on* there being an electron in state i and *none* in state j.

Now we have pointed out that the distribution function f_i (unknown as yet for the present case) may always be interpreted as the average number of electrons in state i, where the average is taken over an ensemble of a great many indistinguishably different systems. In the present situation, however, each state may only have either 1 or 0 electrons in it at any time. So a little thought will show that f_i, the (ensemble) average number of electrons in state i, is identical with the fraction of systems in which the state i is occupied. Similarly, $1 - f_i$ is the fraction of systems in which state i is empty.

Accordingly, if we consider a large ensemble of M systems, the number of transitions taking place within the ensemble from state i to state j in a unit time of observation is just

$$(Mf_i)(1 - f_j)W_{ij}$$

Here Mf_i is thought of as the number of systems in which state i is occupied, $(1 - f_j)$ as the fraction of these systems in which state j is *not* occupied, and W_{ij} as the probability per unit time of a transition, given the other two conditions. Similarly, for transitions from state j to state i, the number taking place per unit time is

$$(Mf_j)(1 - f_i)W_{ji}$$

where of course W_{ji} is the probability per unit time of a transition from state j to state i, given that j is the occupied state and i the empty one. Thus the principle of detailed balancing (applied to the ensemble average transition frequencies) requires that

$$f_i(1 - f_j)W_{ij} = f_j(1 - f_i)W_{ji} \tag{3.24}$$

or

$$\frac{f_i}{1 - f_i} = \left(\frac{f_j}{1 - f_j}\right)\frac{W_{ji}}{W_{ij}} \tag{3.25}$$

Now, as we said before, W_{ji} and W_{ij} are not affected by the exclusion principle, in view of their definitions. Accordingly, using Eq. 3.23 in Eq. 3.25 yields

$$\left(\frac{f_i}{1 - f_i}\right) e^{E_i/kT} = \left(\frac{f_j}{1 - f_j}\right) e^{E_j/kT} \tag{3.26}$$

But notice that the left side of Eq. 3.26 depends *only* on state i, whereas the right side depends *only* on state j. Moreover, the condition of detailed balance must apply for the given system between *any* selected state j, say, and *all* other states i. In other words, Eq. 3.26 must remain true for *all* choices of j and i. The only way in which this can happen is for each side alone to be independent of the state involved. That is,

$$\left(\frac{f_j}{1 - f_j}\right) e^{E_j/kT} = K \text{ (independent of } j) \tag{3.27}$$

In fact, K above may be a function $K(N, T)$ of the number of electrons N in the system considered, and of the temperature T; but it does *not* vary with the dynamical state being considered.

Because the distribution function f_j is the quantity of interest, let us solve for it from Eq. 3.27.

$$f_j = \frac{1}{1 + (1/K)e^{E_j/kT}} \tag{3.28a}$$

For convenience of form, and without any loss of generality whatsoever, it is common practice to rewrite K as:

$$K(N, T) \equiv e^{E_f/kT} \tag{3.28b}$$

which amounts to defining a new function $E_f(N, T)$ as $kT \ln K(N, T)$. Then Eq. 3.28a becomes

$$f_j = \frac{1}{1 + e^{(E_j - E_f)/kT}} \tag{3.29}$$

for the average number of electrons f_j in a state j (including a definite spin specification), corresponding to which is the energy E_j.

Once again, the *form* of Eq. 3.29 is quite independent of any particular properties or arrangements of the allowed quantum states for the particles being considered; but these details of the state arrangement, as well as the number of particles N in the system, will enter when we set out to determine the parameter $E_f(N, T)$ in any given problem by the normalizing requirement

$$\sum_{\text{all } j} f_j = N \tag{3.30}$$

Indeed it is *only* in the determination of E_f that the various problems to which the distribution function in Eq. 3.29 is applicable may differ one from another.

The distribution in Eq. 3.29 is called the *Fermi-Dirac distribution*, and the parameter $E_f(N, T)$ is known as the *Fermi level*. To appreciate the nature of the Fermi-Dirac distribution, consider the plot of f_j given in Fig. 3.2. The symmetry of that plot about the value ($f_j = \frac{1}{2}$, $E_j = E_f$) can be placed in evidence clearly by noting that

$$1 - f_j = \text{average number of electrons missing from state } j$$

$$= \frac{1}{1 + e^{(E_f - E_j)/kT}} \tag{3.31}$$

which is the same function of $E_f - E_j$ that Eq. 3.29 is of $E_j - E_f$.

Observe that at very low temperatures, the exponential function varies extremely rapidly with the difference between E_j and E_f. In the limiting case $T = 0°K$, the value of f_j is *zero* for $E_j > E_f$, and *one* for $E_j < E_f$. This means that E_f at $T = 0°K$ must be somewhere between the highest-energy state occupied by electrons and the lowest empty one, in accordance exactly with the constraints of the exclusion principle. At $T = 0°K$ no electrons can be in states with energies higher than E_f because $f_j = 0$ there; and no states with energies below E_f can be empty because $1 - f_j = 0$ there.

At other temperatures, the Fermi level is the energy at which a state would be occupied in 50% of a large number of identical systems, whereas a state with energy $5kT$ above E_f would be *filled* in only about 0.7% of them, and one with energy $5kT$ below E_f would be *empty* in only about 0.7% of them.

3.5 TEMPERATURE DEPENDENCE OF THE CARRIER CONCENTRATION

3.5.1 *The Diluteness Conditions*

We have argued in Sec. 1.5 that two relations govern the equilibrium concentration of conduction electrons and holes in a

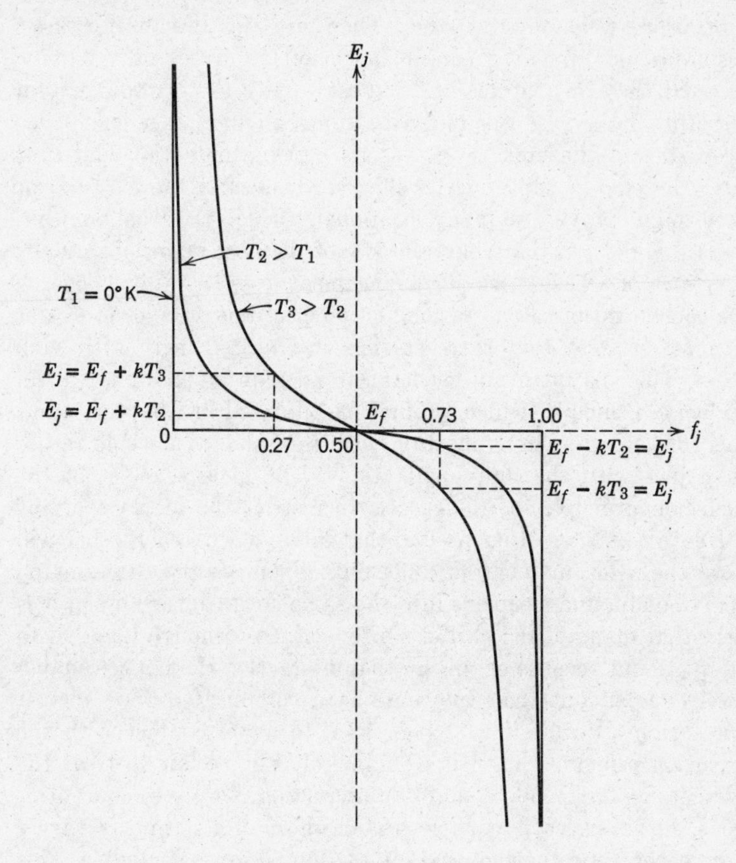

Fig. 3.2. Behavior of the Fermi-Dirac distribution function with energy and temperature.

homogeneous piece of semiconductor; namely, Eqs. 1.18 and 1.19, repeated below for convenience,

$$p_o n_o = n_i^2(T) \tag{3.32a}$$

$$n_o - p_o = N_d - N_a = N \tag{3.32b}$$

Here the intrinsic concentration $n_i(T)$ is characteristic of the host material, independent of the "doping."

It is important to emphasize now that Eq. 3.32*a* was based on the idea that the values of p_o and n_o, and of the net impurity concentration, were not too large. There are two reasons for these assumptions: First, the concentration of impurities must not be so great that their atoms affect substantially the periodic crystal structure. Otherwise, the impurity atoms might change such basic properties of the bulk crystal as its periodic potential, and thus alter the energy gap and the effective masses of the carriers; or they might provide so many locations at which the local environment differs from that characteristic of the host material that the very idea of a basic periodic arrangement loses its utility. Second, the concentration of excess conduction electrons or holes must not become so great that these carriers interact strongly with each other. This is a quantum-mechanical question involving the Pauli exclusion principle, which we first encountered in Chapter 1, but this time it arises in connection with the solid as a whole rather than just with the individual atoms. The general effect of the exclusion principle in this case is to restrict the motions of the ("free") carriers so that no two that interact with each other will have the same momentum and spin. When we try to cram too many conduction electrons into the same space, for example, it is clear that many of the particles often get close enough together to interact, and because of this mutual interaction the particles must surely spread out their momenta in a complicated way, almost like forming "orbits" in a big atom, to avoid conflict with the exclusion principle. Then it would surely not be fair to treat the particles as independent and noninteracting. In dilute concentrations, however, each particle spends most of its time far away from others, and their mutual interaction becomes negligible. This is the most common condition for materials used in transistors and diodes. However, as we have mentioned before, heavily doped material, to which the assumption of independence of carriers does *not* apply, is vitally important in such devices as tunnel diodes and solid-state injection lasers.

We can estimate the concentration of conduction electrons, for instance, at which the exclusion principle would be expected to become important. Use of the effective mass in Eqs. 1.1 and 1.2 of Chapter 1 shows that such a carrier with thermal energy has an uncertainty in position, or "effective size," of

$$\Delta r \cong \frac{h}{\sqrt{3m_e^*kT}} \qquad (3.33)$$

Quantum mechanically, this means that the *de Broglie waves* for such a "particle" extend over a diameter of about this size, so that if the centers of two such particles come closer together than Δr, the waves from the two will interfere and there will be significant quantum interaction. That is, with significant overlapping in space of the particle waves, the two particles may be coming close to having the same quantum state, and must alter their momenta to avoid this condition. So we will expect little effect of the exclusion principle if, on the average, the particles are far apart compared to Δr.

For a uniform spacing between particles of Δr in each direction, there would be one particle in every cube of volume $(\Delta r)^3$, or a number $n_{\text{crit.}}$ per unit volume given by

$$n_{\text{crit.}} \cong \frac{1}{(\Delta r)^3} \cong 5\left(\frac{m_e^* kT}{h^2}\right)^{3/2} \tag{3.34}$$

For the values of m_e^* in Si and Ge, $n_{\text{crit.}}$ is of the order $10^{18} - 10^{19}$ cm^{-3} at room temperature, so we may expect major effects of the exclusion principle on the behavior of the conduction electrons or holes only at concentrations of this order of magnitude or greater.

We wish next to know the values and the temperature dependence of $n_i(T)$ so that we can use Eqs. 3.32*a* and *b* to actually determine the carrier concentrations in different materials over a range of different impurity concentrations and a range of temperatures. But, to be consistent with the limitations on our results up to this point, we will, of course, continue to focus attention mainly on the cases of dilute concentrations of conduction electrons and holes.

3.5.2 *Carrier Concentrations*

Our next problem is to determine explicitly how many of the available valence electrons in a typical unit volume of a homogeneous solid† occupy the various ranges of the allowed states for that volume at any specified thermal equilibrium temperature T. The ranges of allowed states for such valence electrons include those in the conduction band, those in the valence band, those on donors, and those on acceptors.

Actually, we know how many (valence) electrons we are dealing with in a unit volume of the crystal, to be distributed among the aforementioned

† The situation for *inhomogeneous* materials is considered briefly later.

states in the band structure, as a "normalization" condition. If there are M atoms per unit volume of the crystal, of which N_d are substituted donors, and N_a are substituted acceptors, we have 4 valence electrons for each of $M - N_a - N_d$ normal atoms, 5 valence electrons for each of N_d donors and 3 valence electrons for each of N_a acceptors, or $4(M - N_a - N_d) + 5N_d + 3N_a = 4M + (N_d - N_a) = 4M + N$ electrons to deal with. The important point is that this number of electrons must be present at every temperature among the states enumerated above. But using these total numbers is not the best way to insure this "normalization" condition.

The easiest way to keep track of the electrons in the band structure is to note that when they are in the conduction band we call them "conduction electrons," of concentration n_o per unit volume; when they are on acceptors, we say that the acceptor is ionized (negatively charged), of which there may be $N_a^- \leq N_a$ per unit volume; when electrons are *missing* from the valence band, we call the unoccupied levels "holes," of concentration p_0 per unit volume; and, finally, when an extra electron is missing from one of the donors, there is one ionized donor atom (positively charged) left in the lattice, and a possible concentration $N_d^+ \leq N_d$ of these per unit volume. If we start reckoning from the arbitrary "reference condition" in which the valence band is full and all the donors have their extra electrons attached to them, we will have accounted properly at the beginning for all the electrons necessary. Under these arbitrary conditions we have $n_o = 0$, $p_o = 0$, $N_a^- = 0$, and $N_d^+ = 0$. Now, every time we produce a state vacancy that is not present in this reference condition, we must also produce an occupancy that is not initially present. Otherwise, the electrons taken from one situation will not appear in another, and the number of them will not stay the same. This means that all electrons taken from the valence band to produce holes p_o, and all taken from donors to produce ionized donors N_d^+, must appear either in the conduction band as free electrons n_o, or on acceptors to produce ionized acceptors N_a^-. Thus we must have

$$n_o + N_a^- = p_o + N_d^+ \tag{3.35}$$

Often Eq. 3.35 is referred to as a "neutrality condition." It can be interpreted this way, on a macroscopic scale, *only* for *homogeneous* material as follows. Under the reference condition, not only is the entire typical unit volume of the same electrically neutral, but each atomic neighborhood in the crystal is itself electrically neutral. Then as the electrons alter their "positions" in the energy bands, they make some of the microscopic atomic neighborhoods in real space depart from neutrality, but the homogeneous sample must remain neutral in all its typical unit volumes on the *gross* scale. The relation 3.35 expresses this gross neutrality over volumes large enough to contain many particles of each type referred to in it. Notice that Eq. 3.35 is the same as Eq. 3.32*b*, when the impurities are all completely ionized ($N_d^+ = N_d$, $N_a^- = N_a$), which is the only case of further interest to us here. This normalization, or "neutrality" condition, clearly does *not* depend on having a dilute concentration of carriers, whereas we have seen that Eq. 3.32*a* for the $p_o n_o$ product *does* depend on such diluteness. Our statistical considerations cannot give us any more information about the neutrality condition (Eq. 3.32*b*), which indeed must be used as a normalization condition even with the statistical calculation; but they will give us new information about n_i^2 (in Eq. 3.32*a*.)

There are two ways in which we might now proceed to evaluate $n_i(T)$ in Eq. 3.32*a* from a direct calculation involving the Fermi-Dirac distribution function and the known arrangement of available states. One way would be to consider intrinsic material, and simply evaluate n_o and p_o for it, forcing $n_o = p_o = n_i$ by using the normalization condition 3.32*b* for that case. The other procedure is to calculate directly n_o and p_o for any "doped" material (as long as the impurity concentration is dilute enough), and, *independently of condition 3.32b*, show that the *pn*-product is indeed an explicit function of temperature which does *not* involve the impurity concentrations at all. This function will be $n_i^2(T)$.

The second procedure suggested above is no more difficult than the first, and it is much more illuminating. For one thing, it will confirm in a new way our earlier arguments that indeed the *pn*-product is independent of doping; but, more important, it will furnish a direct check on the condition for "diluteness" which we expressed in Eq. 3.34, from physical arguments, as $n_o \ll n_{\text{crit.}}$ for *n*-type material (and, analogously, we would have found $p_o \ll p_{\text{crit.}}$ for *p*-type material).

To proceed, we note first that significant differences in the occupancy of different states, whether in Boltzmann or Fermi-Dirac statistics, occur only if their energy differences are significant *in units of kT*. Recall that kT is about 26×10^{-3} ev at 300°K.

Notice next that the magnitude of momentum differences between adjacent allowed momentum states in the conduction band, for example, is of the order of h/L, which is about $100h$ in mks units for particles in a volume of dimensions $L = 10^{-2}$ m (see Fig. 2.6). Thus, according to Eq. 2.4, the energy difference in kT-units between states with this momentum difference is about

$$\frac{\Delta E}{kT} = \frac{p\Delta p}{m^*kT} = \sqrt{\frac{2(E - E_c)}{m^*}}\,\frac{\Delta p}{kT}$$

or if we consider particles for which the kinetic energy $E - E_c$ is about $kT/2$, we have

$$\frac{\Delta E}{kT} = \frac{\Delta p}{\sqrt{m^*kT}} = \frac{100h}{\sqrt{m^*kT}}$$

At 300°K for electrons this works out to be a fraction 10^{-6}. Hence there is a tremendous number of allowed momentum states for which the energy in kT-units, and therefore the probability of occupancy, is virtually the same. And this is in addition to the multiplicity of states we previously mentioned, with *exactly* the same energy, for which the *direction* of motion is simply different.

Accordingly, if we consider a three-dimensional differential increment of momentum, $dp_x\, dp_y\, dp_z$, which can be represented as a box in the momentum space of Fig. 2.6*b* small enough so that the distribution function f_i does not change appreciably throughout it (because the energy does not), there is still a very large number of "allowed states" within that interval. In other words, the microscopic momentum volume shown in Fig. 2.6*b*, in which only two electron states can fit, is *much* smaller than even the *macroscopic differential* volume $dp_x\, dp_y\, dp_z$ in which f_i is substantially constant. Specifically, if we consider a unit volume of the solid ($V = 1$, in whatever units are being used, in Fig. 2.6), no more than *one* electron can have a momentum within *one-half* of the box shown in Fig. 2.6*b*, and this volume in momentum space is $h^3/2$. Hence, the number of allowed electron states within the macroscopic differential volume $dp_x\, dp_y\, dp_z$ is

$$\text{number of states} = \frac{2dp_x\, dp_y\, dp_z}{h^3} \quad \text{(per unit volume of crystal)} \qquad (3.36)$$

Because the average number of electrons occupying each state in Eq. 3.36 is practically the same, given by

$$f = \frac{1}{1 + e^{(E-E_f)/kT}}$$

with

$$E = E_c + \frac{1}{2m_e^*}(p_x^2 + p_y^2 + p_z^2) \qquad (3.37)$$

for the conduction band, the average total number of electrons occupying all the states in this band (i.e., the total number n of conduction electrons per unit volume) is just the product $[(2/h^3)\, dp_x\, dp_y\, dp_z] f$ added up for all values of momentum in all directions in the conduction band. That is,

$$n_0 = \int_{-\infty}^{+\infty}\int_{-\infty}^{+\infty}\int_{-\infty}^{+\infty} \frac{2}{h^3} \frac{dp_x\, dp_y\, dp_z}{1 + e^{(E_c-E_f)/kT}\, e^{p_x^2/2m_e^*kT}\, e^{p_y^2/2m_e^*kT}\, e^{p_z^2/2m_e^*kT}} \qquad (3.38)$$

Now the condition of diluteness we have mentioned before implies that each state in the conduction band is sparsely occupied. That is, the average number of electrons in each state i is small, or the fraction of time during which any such state is occupied by an electron is very small. Thus f_i is much smaller than 1, and the 1 in the denominator of Eq. 3.38 must be negligible. Such a condition will occur for *all* states in Eq. 3.38 only if it occurs for $p_x = p_y = p_z = 0$; and this requires that

$$E_c - E_f \gg kT \tag{3.39}$$

Of course, we do not yet know the location of E_f; it must be determined by the normalization condition later. But let us proceed on the diluteness assumption that Eq. 3.39 is satisfied and check back later on this assumption. The distribution function becomes Boltzmann in form (which is just what we would expect for dilute populations in which the exclusion principle has little effect). Then Eq. 3.38 reads:

$$\frac{h^3}{2} e^{(E_c - E_f)/kT} n_o =$$

$$\int_{-\infty}^{\infty} \int_{-\infty}^{\infty} \int_{-\infty}^{\infty} e^{-p_x^2/2m_e^*kT} e^{-p_y^2/2m_e^*kT} e^{-p_z^2/2m_e^*kT} dp_x\, dp_y\, dp_z \tag{3.40}$$

The three integrals in Eq. 3.40 are all identical. One such integral, for example, is

$$\int_{-\infty}^{\infty} [\exp(-p_x^2/2m_e^*kT)]\, dp_x = \sqrt{2m_e^*kT} \int_{-\infty}^{\infty} e^{-y^2}\, dy \tag{3.41}$$

where the normalized variable has been used

$$y \equiv \frac{p_x}{\sqrt{2m_e^*kT}} \tag{3.42}$$

Because the integrand is an even function of y, we get from Eqs. 3.17 and 3.18 for the case $n = 0$ the result

$$\int_{-\infty}^{\infty} e^{-y^2}\, dy = 2 \int_0^{\infty} e^{-y^2}\, dy = \sqrt{\pi} \tag{3.43}$$

and Eq. 3.41 becomes

$$\int_{-\infty}^{\infty} [\exp(-p_x^2/2m_h^*kT)] dp_x = \sqrt{2\pi m_e^*kT} \tag{3.44}$$

So, considering all three integrations, Eq. 3.40 gives for n_0,

$$n_o = 2\left(\frac{2\pi m_e^*kT}{h^2}\right)^{3/2} e^{-(E_c - E_f)/kT} \tag{3.45}$$

If we put

$$N_c \equiv 2\left(\frac{2\pi m_e^* kT}{h^2}\right)^{3/2} \tag{3.46}$$

Eq. 3.45 becomes

$$n_o = N_c e^{-(E_c - E_f)/kT} \tag{3.47}$$

Evidently, if $E_c - E_f \ll kT$, we have $n_o \ll N_c$, and conversely. A comparison of N_c in Eq. 3.46 with $n_{\text{crit.}}$ in Eq. 3.34 shows that they are essentially the same (except for a numerical constant of the order of 6) and justifies our arguments about diluteness.

It is worth pointing out that in a metal, or a degenerate semiconductor, the concentration of electrons free to conduct is comparable to, or much greater than, $n_{\text{crit.}}$; so even though Eq. 3.38 may still be applicable, the 1 in the denominator is *not* negligible. The integration involved is then more difficult to carry out (it has been tabulated), and the resulting electronic behavior is more complicated. The fortuitous diluteness of carrier concentration in some common semiconductors, like the Ge and Si used in transistors, is the main feature that makes the analysis of their behavior amenable to the kind of simplified treatment we have been presenting and will continue here.

To compute p_o, the number of holes, we must find the number of electron states in the valence band which are *not* occupied. The number of states per unit volume, per momentum increment, is the same for the valence band as for the conduction band (Eq. 3.36), because we have accepted the fact that a hole behaves as a nearly classical particle similar to a conduction electron (but with effective mass m_h^* instead of m_e^*). In this case, the distribution for *vacant* states (i.e., the average number of electrons *missing* from a given state) is

$$1 - f = \frac{1}{1 + e^{(E_f - E)/kT}}$$

and here, in the valence band, according to Eq. 2.6 we have

$$E = E_v - \frac{1}{2m_h^*}(p_x^2 + p_y^2 + p_z^2) \tag{3.48}$$

The integration must now be of $(1 - f)[(2/h^3)dp_x\,dp_y\,dp_z]$ over all momenta in the valence band, and if we may again assume a dilute population (of holes, this time),

$$E_f - E_v \gg kT \tag{3.49}$$

we will find

$$p_o = \int_{-\infty}^{\infty}\int_{-\infty}^{\infty}\int_{-\infty}^{\infty} e^{-(E_f-E_v)/kT}\, e^{-(p_x^2+p_y^2+p_z^2)/2m_h{}^*kT}\, dp_x\, dp_y\, dp_z \quad (3.50)$$

or by using the same integration as before,

$$p_o = N_v e^{-(E_f-E_v)/kT} \quad (3.51)$$

where

$$N_v = 2\left(\frac{2\pi m_h{}^* kT}{h^2}\right)^{3/2} \quad (3.52)$$

Again, N_v has the significance of a $p_{\text{crit.}}$. The values of N_c and N_v depend to some extent on the temperature ($T^{3/2}$), but at around room values (300°K) they are of the order of 10^{19} cm^{-3} in Si and Ge. Evidently, at a given temperature the difference between values of these parameters in different materials depends on the effective masses.

Observe that the diluteness conditions Eqs. 3.39 and 3.49 taken together imply that the energy gap is large in kT-units:

$$E_c - E_v = E_g \gg kT \quad (3.53)$$

which *must be* true for all transistor materials that are going to be useful in devices at temperature T.

3.5.3 *The p_on_o-Product*

We are now in a position to observe once again (this time from Eqs. 3.47 and 3.51) that the product p_on_o is independent of the impurity concentration. That is,

$$\begin{aligned} p_on_o &= N_cN_v e^{-(E_c-E_v)/kT} = N_cN_v e^{-E_g/kT} \\ &= 32\left(\frac{\pi km}{h^2}\right)^3 \left(\frac{m_e{}^* m_h{}^*}{m^2}\right)^{3/2} T^3 e^{-E_g/kT} = n_i^2(T) \end{aligned} \quad (3.54a)$$

where we have used m, the true mass of an electron outside the solid, as a convenient reference base for the effective masses. We have in Eq. 3.54*a* not only validated Eq. 3.32*a* again, but also evaluated $n_i^2(T)$.

Unfortunately, even the *form* of $n_i^2(T)$, namely,

$$n_i^2(T) = B(m_e{}^* m_h{}^*)^{3/2} T^3 e^{-E_g/kT} \quad (3.54b)$$

cannot be derived without considering the quantum effects of both the uncertainty principle and the Pauli exclusion principle (consideration of these appears in Sec. 3.5.2). However, a plausibility argument may be used to show the meaning of Eq. 3.54*b*, even though it furnishes no proof.

Consider intrinsic material. To produce a hole-electron pair *at rest* requires an ionization energy E_g. Assign *arbitrarily* an amount E_o of this to the hole, and $E_g - E_o$ to the electron, effectively as "potential" energies. But the conduction electrons and holes are *not* at rest—they have thermal momenta and, accordingly, kinetic energies. So the number of conduction electrons (n_i in this case) must include all those with all possible momenta; namely, from Boltzmann statistics for the dilute population,

$$n_i = C \int_{-\infty}^{\infty} \int_{-\infty}^{\infty} \int_{-\infty}^{\infty} \exp\left[-\frac{(E_g - E_o)}{kT}\right] \exp\left[-\frac{(p_x^2 + p_y^2 + p_z^2)}{2m_e^*kT}\right] dp_x\, dp_y\, dp_z$$

From this, using the integrals in Eqs. 3.17 and 3.18 for the case $n = 0$, we have

$$n_i = C'(m_e^*kT)^{3/2}e^{-(E_g - E_o)/kT}$$

where the numerical factors from the integration are absorbed in C'.

Similarly, the number of holes having all momenta (n_i again) must be given by

$$n_i = D'(m_h^*kT)^{3/2}e^{-E_o/kT}$$

Then we find that our high-handed (and very questionable) choice of E_o does not matter in the product of the two results above, namely,

$$n_i^2 = C'D'(m_e^*m_h^*)^{3/2}(kT)^3e^{-E_g/kT}$$

If we will accept on faith the (not obvious) fact that the coefficient $C'D'$ does not depend on T in this case, we will have accounted in this manner for the form of Eq. 3.54*b* and will perhaps have gained a little insight into the interpretation of the various factors that appear.

To use the Eq. 3.54 results effectively, however, we must remember that the energy gap changes with temperature because of changes in the periodic crystal lattice. Increasing temperature decreases the gap for Ge and Si, as listed in Table 1.1. In Eq. 3.54, the value of E_g must be chosen accordingly. To the extent that this variation is linear in T, as indicated in Table 1.1, we may simplify our results by using the assumption that

$$E_g(T) = E_{go} - \beta T \tag{3.55}$$

where β is the rate of decrease of gap with increasing temperature, and E_{go} is the *extrapolated* value of the gap at $T = 0°K$. E_{go} is not the *actual* gap at $T = 0°K$, because the linear form in Eq. 3.55 is grossly inaccurate at low temperatures. On the basis of Eq. 3.55, however, Eq. 3.54 becomes

$$p_o n_o = n_i^2(T) = \left[32\left(\frac{\pi k m}{h^2}\right)^3 \left(\frac{m_e^* m_h^*}{m^2}\right)^{3/2} e^{\beta/k}\right] T^3 e^{-E_{go}/kT} \tag{3.56}$$

Introduction of numerical values of m_e^* and m_h^* from Table 1.0, and also of the values of E_{go} and β given in Table 1.1 (in ev), should give equations that agree with the following measured results,† applicable for temperatures above about 50°K:

$$p_o n_o = n_i^2(T) = 3.10 \times 10^{32} T^3 e^{-9100/T} \text{ cm}^{-6} \text{ for Ge} \tag{3.57a}$$

$$p_o n_o = n_i^2(T) = 1.50 \times 10^{33} T^3 e^{-14000/T} \text{ cm}^{-6} \text{ for Si} \tag{3.57b}$$

We may not insist too strongly on the relationship between the numerical values given in Eqs. 3.57 and the formula 3.56, however, because independent measurement of the effective masses that appear in Eq. 3.56 is quite difficult to perform. In general, it is not even possible to do so at the high temperatures for which Eqs. 3.57*a* and *b* apply. Recall also that the temperature range for Eq. 3.56 is restricted on the low end by failure of Eq. 3.55 at low temperatures (among other difficulties). Nevertheless, for its general form, the square root of Eqs. 3.54 and 3.56 yields

$$\begin{aligned} n_i(T) &= 5.66 \left(\frac{\pi k m}{h^2}\right)^{3/2} \left(\frac{m_e^* m_h^*}{m^2}\right)^{3/4} T^{3/2} e^{-E_g/2kT} \\ &= \left[5.66 \left(\frac{\pi k m}{h^2}\right)^{3/2} \left(\frac{m_e^* m_h^*}{m^2}\right)^{3/4} e^{\beta/2k}\right] T^{3/2} e^{-E_{go}/2kT} \end{aligned} \tag{3.58}$$

† E. M. Conwell, *Proc. IRE*, **46** (June 1958), pp. 1281–1300.

The corresponding experimental results from Eqs. 3.57 are

$$n_i(T) = 1.76 \times 10^{16} T^{3/2} e^{-4550/T} \text{ cm}^{-3} \qquad \text{for Ge} \qquad (3.59a)$$

$$n_i(T) = 3.88 \times 10^{16} T^{3/2} e^{-7000/T} \text{ cm}^{-3} \qquad \text{for Si} \qquad (3.59b)$$

and these are plotted in Figs. 1.9 and 1.10.

At $T = 300°\text{K}$, the foregoing data yield

$$n_i(300°\text{K}) = 2.4 \times 10^{13} \text{ cm}^{-3} \qquad \text{for Ge}$$
$$= 1.5 \times 10^{10} \text{ cm}^{-3} \qquad \text{for Si}$$

Notice that in the temperature range from 100°K to 1000°K, which is easily achievable in most laboratories, the term $T^{3/2}$ in Eqs. 3.59 increases by a factor of about 32. This seems a large amount until we observe that the exponentials increase by approximately the factors $e^{41} \approx 6 \times 10^{17}$ for Ge and $e^{63} \approx 2 \times 10^{27}$ for Si. Thus changes in the $T^{3/2}$ term are, by and large, not very important; $n_i(T)$ is nearly an exponential function of $(1/T)$ over a considerable temperature range (refer to Figs. 1.9 and 1.10).

Incidentally, the melting points of Ge and Si are, respectively, 1210°K and 1693°K, and at these temperatures Eqs. 3.59 predict

$$n_{i,\text{melt}}(1210°\text{K}) = 1.7 \times 10^{19} \text{ cm}^{-3} \qquad \text{for Ge}$$
$$n_{i,\text{melt}}(1693°\text{K}) = 4.4 \times 10^{19} \text{ cm}^{-3} \qquad \text{for Si}$$

Such "predictions" are, of course, to be taken only for rough numerical orientation purposes, inasmuch as Eqs. 3.59 came from measurements made only up to about 1000°K. There are about 5×10^{22} atoms/cm^3 in the Si or Ge lattice, and much of the energy that holds the lattice together comes from the electron bonds. We therefore expect a significant fraction of these atoms to have at least one broken bond when melting occurs. The result above is suggestive of this fact, but it would be no surprise to learn that, quite near the melting point, the behavior of the energy gap with temperature is likely to depart from the simple form used in Eq. 3.55.

In any case, the number of "intrinsic" conduction electrons or holes (n_i each) which are present in these semiconductor materials without benefit of added impurities, is approximately an exponential function of the form exp $(-E_{go}/2kT)$ over a very wide temperature range.

For small temperature changes ΔT about a given value T_1°K we can find an even simpler form of the above result by using the approximation

$$\frac{E_{go}}{2k(T_1 + \Delta T)} = \frac{E_{go}}{2kT_1(1 + \Delta T/T_1)} \approx \frac{E_{go}}{2kT_1}\left(1 - \frac{\Delta T}{T_1}\right)$$

provided $\Delta T/T_1 \ll 1$. Thus we may rewrite Eq. 3.58 in the approximate form

$$n_i(\Delta T) \approx n_i(T_1) \exp \frac{E_{go}\Delta T}{2kT_1^2}; \quad \text{for } \frac{\Delta T}{T_1} \ll 1 \qquad (3.60)$$

which is a *direct* exponential in the temperature increment ΔT. One convenient choice for T_1 is 273°K, so that ΔT becomes the centigrade temperature directly. With that choice, the exponent in Eq. 3.60 becomes 6.1×10^{-2} per °C for Ge and 9.4×10^{-2} per °C for Si, which means that n_i increases with temperature in the neighborhood of 0°C at the fantastic rate of about 6% per °C for Ge and 9% per °C for Si! We shall see as we proceed that this intrinsic property of semiconductors places very important environmental constraints upon semiconductor devices.

3.5.4 *Ionization of Impurities*

We have almost finished the background required to justify completely Eqs. 3.32. The remaining point to be made concerns a check on the fact that the donors and acceptors are almost 100% ionized at normal temperatures. As we have said, from the elementary ideas it seems reasonable that, with the average thermal energy of about $kT = 26$ millielectron volts at room temperature, it would not be difficult to ionize a good many donors (about 10 to 50 millielectron volts are needed to do so in Ge and Si), but it would be relatively hard to produce many hole-electron pairs (600 to 1000 millielectron volts are needed). Yet the comparison of 26 with 10 to 50 does not seem to suggest offhand the tremendously high percentage of ionization of donors we have been implying. So, let us examine the situation more closely.

Without any impurities, we have seen that about 10^{13} intrinsic hole-electron pairs are present per cubic centimeter in Ge at room temperature, and about 10^{10} in Si. On the other hand, N_c and N_v are about 10^{19}cm^{-3} in both materials. One way of looking at the meaning of N_c and N_v comes from Eqs. 3.47 and 3.51. Interpreted as Boltzmann factors between energies E_c and E_f for n, and E_f and E_v for p, these equations give N_c the significance of an "effective concentration of available states" in the conduction band, all seeming to be located exactly at the band edge E_c, and N_v similarly an effective concentration of available "hole" states, all at energy E_v. So with the numerical values just noted, the "available" electron states in the conduction band (or hole states in the valence band) are seen to be really almost empty under intrinsic conditions. Less than one state per million available is filled! This is simply another way of saying that $(E_c - E_f)/kT > \ln 10^6 = 14 \gg 1$, for the intrinsic material.

If, for example, donors are now added in the amount $N_d = 10^{16}\text{cm}^{-3}$, which is the "geometric mean" $(= \sqrt{n_i N_c})$ between $n_i = 10^{13}\ \text{cm}^{-3}$ and $N_c = 10^{19}\ \text{cm}^{-3}$ for Ge, our previous reasoning indicates that these donors should all be substantially ionized and contribute their electrons to the conduction band (i.e., make their electrons "free" for conduction purposes). This suggests that n must be approximately $10^{16}\ \text{cm}^{-3}$. On this basis, $n = 10^{-3}N_c$, so that the available states in the conduction band are *still* not very well filled. The exponential factor in Eq. 3.47 is, in fact, equal to 10^{-3}, which means $(E_c - E_f)/kT = 3 \ln 10 = 6.91$. This is still large enough to justify the "diluteness" approximations we have been making. But now we recall that for donors, $E_c - E_d$ is in the range 0.01 to 0.05 ev, which amounts to, say, about kT at room temperature. Thus $(E_d - E_f)/kT = (E_c - E_f)/kT - (E_c - E_d)/kT$, or about 6. Although on the basis of our discussions here we would use the Fermi-Dirac distribution, Eq. 3.29, to evaluate the fraction occupied of the N_d discrete donor states per unit volume, the 1 in the denominator is entirely negligible in this case and

$$f_d \cong e^{-(E_d - E_f)/kT} = e^{-6} = 1/400$$

which is only $\frac{1}{4}\%$! The donors *are* well over 99% ionized; but we notice from the analysis that the reason for it is not so much that their energies are close to the conduction band, as that there are so few donor states $N_d(=n)$ compared to N_c! The physical interpretation of this fact is that there are so many ways an electron can be "free" in the conduction band, and so (relatively) few donor atoms per unit volume of crystal, that an electron once released from a donor has practically no chance of finding another donor in its wanderings. So almost all donor electrons become, and remain, "free" in the conduction band.

3.6 SOME CONSEQUENCES OF THE EQUILIBRIUM CONDITION

3.6.1 *Nonuniform Material*

An interesting circumstance arises when we consider a piece of semiconductor material in which the concentration of impurities is not uniform everywhere. Of course, we shall be very much concerned with such matters later in connection with semiconductor devices, nearly all of which have *pn* junctions in them, so we shall not attempt to treat the situation in great detail now.

For our present needs, let us imagine a long bar containing a graded concentration of donors and acceptors, so that donors dominate at the left end and acceptors at the right end (Fig. 3.3*a*). We wish to reason only *qualitatively* about this arrangement, in order to appreciate the general form of the rest of Fig. 3.3—not its details. The impurity grading is supposed to be gradual enough to be considered as negligible over distances of a substantial number of lattice periods, so that the band structure makes sense over all distances on the laboratory scale. In particular, the energy gap will be the same everywhere, as will the impurity ionization energies. Thus, if the temperature is high enough to ionize all the impurities, there is grading of *immobile* charge density qN on the impurity atoms from a substantial positive value on the donor end at the left, through zero where the donor and acceptor concentrations are equal, to a substantial negative value at the acceptor end to the right (Fig. 3.3*b*).

Correspondingly, on first thought it would seem that there ought to be a neutralizing cloud of mobile carriers at each point, comprised almost entirely of free electrons at the left and holes at the right. Were this the case, however, the increase in concentration of free electrons from right to left, and of holes from left to right, would produce a diffusion current of both carriers toward the center. This action would leave an excess (unneutralized) concentration of fixed positive donor charge on the left and fixed negative acceptor charge on the right, in addition to building up an excess of mobile holes on the left and conduction electrons on the right (Figs. 3.3*c* and *d*). The net charge density ρ that would result (Fig. 3.3*e*) is a "dipole" distribution that is positive on the left and negative on the right, producing an electric field E_x and an electrostatic potential distribution $\Psi(x)$ that would tend to *oppose* the diffusion discussed above.

Accordingly, the graded regions of the bar are *not* in fact neutral and there *is* a distribution of electrostatic field and potential just sufficient to maintain a steady state of affairs (Fig. 3.3*f*). Can we say any more about how this electrostatic potential affects the distribution of conduction electrons and holes?

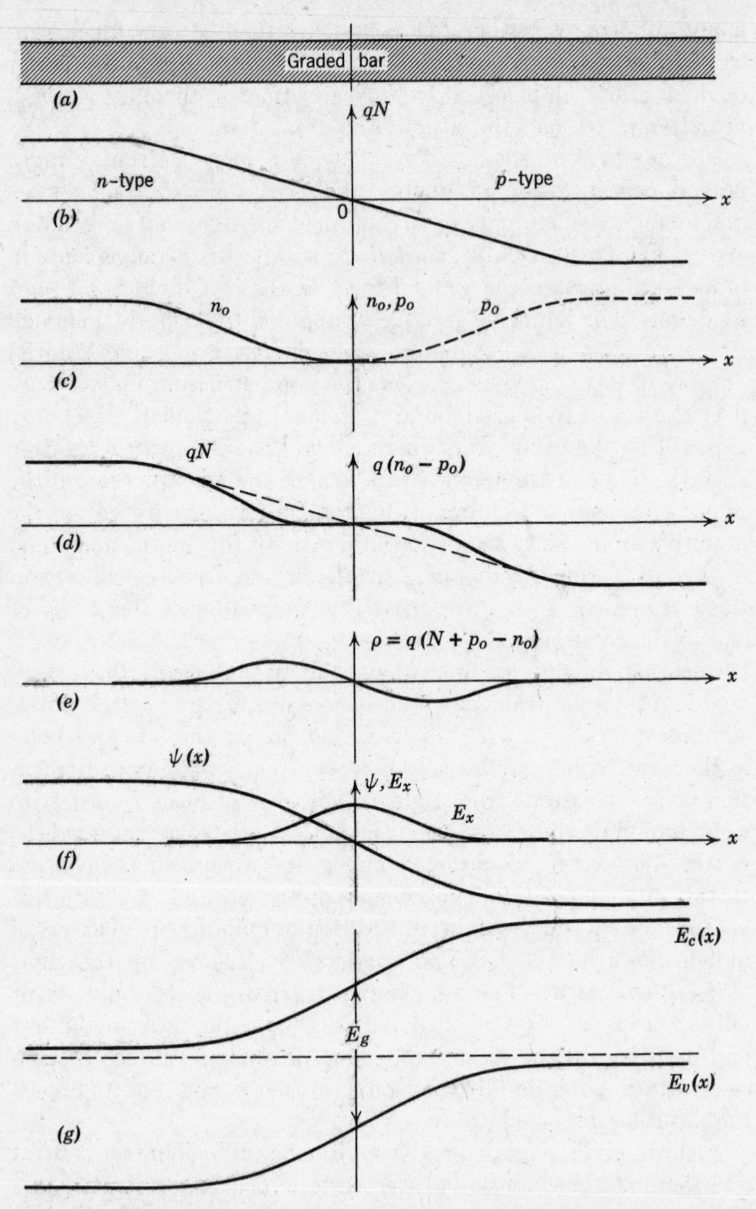

Fig. 3.3. Equilibrium properties of an inhomogeneous bar. (*a*) Structure. (*b*) Net fixed charge. (*c*) Carrier concentrations. (*d*) Net mobile charge. (*e*) Net total charge. (*f*) Potential ψ and field E_x; (*g*) Electron energy-band diagram.

We have already made it clear that in cases of interest to us the free carriers behave like ordinary charged particles of effective mass m^*, in response to relatively weak forces which tend to move them with respect to the crystal lattice. Such charged particles will behave in an electrostatic potential just exactly as massive gas particles will behave under a gravitational potential. In particular, the potential energy of holes in the electrostatic potential $\Psi(x)$ is just $+q\Psi(x)$, and that of free electrons is just $-q\Psi(x)$. Thus, if we elect to choose our space origin $x = 0$ at the place where the semiconductor is intrinsic (i.e., where $n_o = p_o = n_i$) and to choose $\Psi(0) = 0$ there too, we know from previous considerations that for the noninteracting holes in thermal equilibrium a "Boltzmann factor" like Eq. 3.9 or 3.14 will apply:

$$p_o(x) = n_i e^{-q\psi(x)/kT} \tag{3.61}$$

and similarly for the free electrons:

$$n_o(x) = n_i e^{+q\psi(x)/kT} \tag{3.62}$$

Observe that Eqs. 3.61 and 3.62 are consistent *at each point* with the requirement that in thermal equilibrium

$$p_o(x) n_o(x) = n_i^2(T)$$

for all reasonable impurity concentrations in material of concern to us here. We shall have considerable need for the relations in Eqs. 3.61 and 3.62 in our work on *pn* junctions.

Occasionally it is useful to extend the electron energy-band diagrams (similar to Fig. 2.5) to describe nonuniform materials in thermal equilibrium. The basis of this extension was discussed partly in Sec. 2.2.2. Now all we need to add is that the conduction-band edge energy E_c (in joules) represents the energy of a free electron without any kinetic energy! Thus $E_c = -q\Psi(x)$. That is, E_c varies with position as the negative of the electrostatic potential. Similarly, an electron *missing* from the valence-band edge E_v (joules) represents a hole without any kinetic energy, and one of these can always be produced by raising an electron from E_v to E_c—an energy increase of the gap amount, E_g. Thus E_v must also be a function of position $E_v(x) = E_c(x) - E_g = -[q\Psi(x) + E_g]$. The energy gap is constant throughout, as long as the underlying semiconductor is everywhere the same, regardless of the variation over reasonably wide limits of the impurity concentration. A sketch of the "energy-band diagram" showing these features appears in Fig. 3.3*g*. The dashed line at fixed energy simply reminds us that $E_c(x)$ and $E_v(x)$ must be measured with respect to one, and only one, reference potential energy for the whole system.†

If the ends of the bar in Fig. 3.3 are composed of uniform material, characterized by known concentrations $n_{-\infty} = N_d$ free electrons per cm³ in the *n*-type material on the left, and $p_{+\infty} = N_a$ holes per cm³ in the *p*-type material on the right, it is possible to determine the difference in electrostatic potential between the ends by solving Eqs. 3.61 and 3.62 for the potentials. Thus,

$$\begin{aligned} \Psi(-\infty) &= (kT/q)\ln(n_{-\infty}/n_i) \cong (kT/q)\ln(N_d/n_i) \\ \Psi(+\infty) &= -(kT/q)\ln(p_{+\infty}/n_i) \cong -(kT/q)\ln(N_a/n_i) \end{aligned} \tag{3.63}$$

and therefore

$$\begin{aligned} \Delta\Psi = \Psi(-\infty) - \Psi(+\infty) &= (kT/q)\ln(n_{-\infty}p_{+\infty}/n_i^2) \\ &\cong (kT/q)\ln(N_dN_a/n_i^2) \end{aligned} \tag{3.64}$$

This electrostatic potential is not only required to produce a deadlock with the diffusive tendencies of the free carriers, as we said before, *but it must do so separately for conduction electrons and holes.* It is obvious that the total current in the open-circuited bar has to be zero; but it is less obvious that the currents of holes and free electrons have to be zero *separately.*

Here we come again to a case of the principle of detailed balance, in which the opposing processes of diffusion and drift under thermal equilibrium conditions must self-balance for each type of carrier separately. If this were not so, a net hole current one way would have to be cancelled by a net conduction-electron current the other way—just to get the open-circuit conditions met. But holes and free electrons carrying opposing currents are actually moving in the *same* direction! Hence the cancellation of currents implies a steady migration of pairs from one end of the bar to the other. Either the situation is not a steady state, or pairs are steadily

† For those familiar with the Fermi-Dirac statistics, it may be helpful to point out that the level shown is actually the Fermi level for the system. The Fermi level is independent of position because there is only *one* such parameter in the distribution for any system—homogeneous or not.

being generated on one end and moving to the other end to recombine. Such a process would draw energy out of one end of the bar, and give it up to the other (assuming the bar was isolated thermally from its surroundings). One end gets cold, while the other end gets hot, all by spontaneous action in "equilibrium." Not possible by the second law of thermodynamics! So the currents must self-cancel individually.

Incidentally, it is interesting to show that no current can be drawn from the potential difference $\Delta\Psi$, because to do so would also violate the second law by heating up the "load" used and cooling off the "source" bar spontaneously, under thermal equilibrium conditions.

3.6.2 *Drift, Diffusion, and the Einstein Relation*

In the last section we noted that the hole and conduction-electron currents were necessarily zero separately in thermal equilibrium, even in a graded bar (Fig. 3.3). In Chapter 1 we developed the fact that the mobile carriers produced current by both the conduction (or drift) mechanism and the diffusion mechanism. Taking holes, for example, the total hole current density (amperes/m^2) would be the sum of the drift and diffusion components:

$$\mathbf{J}_h = q\mu_h p\mathbf{E} - qD_h\nabla p \tag{3.65a}$$

and for conduction electrons

$$\mathbf{J}_e = q\mu_e n\mathbf{E} + qD_e\nabla n \tag{3.65b}$$

Notice that the drift terms in both equations above have the same algebraic sign because the electric field produces current in the *same* direction for oppositely charged particles; but the diffusion terms have *opposite* algebraic signs when written as amperes (rather than as particle flows), because a given concentration gradient would cause both kinds of particles to diffuse in the *same* direction, and therefore to carry electric current in *opposite* directions on account of the opposite charges.

Equations 3.65*a* and *b* will be very important to us in nonequilibrium situations when we take up the internal electronics of transistors and related devices, but at the moment we wish only to apply them under thermal equilibrium conditions. In that case, $\mathbf{J}_h = 0$. So from Eq. 3.65*a* we have

$$\mathbf{J}_h = q\mu_h p_o \mathbf{E} - q D_h \nabla p_o = 0 \tag{3.66}$$

But the electrostatic potential Ψ is related to E as follows

$$\mathbf{E} = -\nabla\Psi \tag{3.67}$$

which may be used in Eq. 3.66 to obtain

$$\nabla p_o / p_o = -(\mu_h / D_h)\nabla\Psi$$

or

$$\nabla\left(\ln p_o + \frac{\mu_h}{D_h}\Psi\right) = 0$$

or

$$\ln p_o + \frac{\mu_h}{D_h}\Psi = \text{const.}$$

or, finally, if C is a constant independent of position

$$p_o = Ce^{-(\mu_h/D_h)\psi} \tag{3.68a}$$

Similarly for the conduction electrons

$$n_o = C'e^{(\mu_e/D_e)\psi} \tag{3.68b}$$

where C' is independent of position.

Comparison of Eqs. 3.68 with Eqs. 3.61 and 3.62 shows that we must have

$$D_h/\mu_h = D_e/\mu_e = kT/q \tag{3.69}$$

The mobility and the diffusion constants of the carriers *in a given material* are not independent of one another, but are connected by Eq. 3.69, known as the *Einstein relation.* This relation will simplify our discussions of the current-flow problems to be taken up in the next chapter. It will also be used repeatedly in the treatment of diodes and transistors. Note, however, that the validity of the Einstein relation depends upon the applicability of the Boltzmann factors 3.61 and 3.62, the approximations for which have been discussed in detail previously.

3.6.3 *Summary of Relations Involving Carrier Concentrations*

It is well at this point to re-emphasize the domains of applicability of some of our principal conclusions.

The normalizing or macroscopic neutrality relationship, written on a per-unit-volume basis

$$n_o - p_o = N_d - N_a = N \tag{1.19}$$

has been presented as a property only of *homogeneous* material, in thermal equilibrium. It is certainly *not* true on such a point-by-point basis for *inhomogeneous* material. For the latter, of course, the difference between the *total* number of conduction electrons and holes in an *entire* sample must be accounted for by a difference between the *total* number of ionized donors and acceptors in the sample; but for inhomogeneous material the net excess of one need not (and generally does not) appear at the same place in space as the net excess of the other (see Fig. 3.3). The additional relationships needed for a full discussion of the variation of p and n with position in *in*homogeneous material will be developed in the next chapter, and the entire situation is explored further elsewhere.†

The *pn*-product relation

$$p_o n_o = n_i^2(T) \tag{3.32a}$$

holds in thermal equilibrium *only*, but it holds for homogeneous and (at each point) for inhomogeneous material as well—provided only that the doping is not too concentrated. The same is true of the related results

$$p_o = n_i e^{-q\psi/kT} \tag{3.61}$$

$$n_o = n_i e^{+q\psi/kT} \tag{3.62}$$

The flow equations

$$\mathbf{J}_h = q\mu_h p\mathbf{E} - qD_h \nabla p \tag{3.40a}$$

$$\mathbf{J}_e = q\mu_e n\mathbf{E} + qD_e \nabla n \tag{3.40b}$$

apply especially for first-order departures from thermal equilibrium. They are, in fact, the *simplest nonequilibrium flow relations* we can have, because they are linear in the driving "forces" **E** and ∇p or ∇n. They will form the basis of much of our discussion

† P. E. Gray, D. DeWitt, A. R. Boothroyd, and J. F. Gibbons, *Physical Electronics and Circuit Models of Transistors*, John Wiley and Sons, New York (1964), SEEC Vol. 2, hereafter referred to simply as PEM.

in the next chapter. Applied to situations at thermal equilibrium, however, $\mathbf{J}_e \equiv \mathbf{J}_h \equiv 0$, and the *only* conclusions forthcoming from these equations are the Einstein relations $D_{e,h} = \dfrac{kT}{q}\mu_{e,h}$ and a repeat of Eqs. 3.61 and 3.62.

PROBLEMS

P3.1 An uncharged particle of mass m is constrained to move along the x-axis without friction, and is attracted to the origin by a spring force $f = -Kx$.

(*a*) Classically, show that the particle naturally executes a simple harmonic motion of radian frequency $\omega_o = \sqrt{K/m}$, and an amplitude proportional to the square-root of its total energy E, which may have any positive value desired.

(*b*) By considering an ensemble of N identical oscillators in thermal equilibrium at temperature T, obtain the classical distribution density function for the number of oscillators having position in the range between x and $x + dx$, *and* momentum in the range between p_x and $p_x + dp_x$. To do this, make use for the oscillator of the relation between energy, momentum, and position.

(*c*) Normalize the distribution of part (*b*) properly.

(*d*) Show that the average kinetic and potential energies for such an oscillator are each equal to $\frac{1}{2}kT$.

(*e*)† Quantum-mechanically, it is a well-known result that a harmonic oscillator of the type described above may only take on discrete states, each labelled with an integer n and having an unique value E_n of total energy, given by the expression $E_n = (n + \frac{1}{2})h\omega_0/2\pi$, for $n = 0,1,2, \ldots$. Determine the distribution *function* (*not density* in this case), giving the number of oscillators in state n, properly normalized for the quantum-mechanical case. Show under what circumstances it may be reduced to the classical distribution *density* found in part (*b*).

P3.2 (*a*) Find the *average* kinetic energy of a monatomic ideal gas molecule in thermal equilibrium at temperature T.

(*b*) Find the *most probable* kinetic energy, and plot versus E the distribution density for the number of molecules with kinetic energy between E and $E + dE$.

(*c*) Find the average *speed*.

(*d*) Make use of your result (*b*) above, and the model proposed in P2.4*b*, to find *approximately* the wavelength of the *most intense* recombination radiation you would expect to observe from a simple semiconductor at temperature T. Make a sketch of the form of the spectral "line" you expect from such radiation (you may wish to use graphical methods for this).

† These parts require more than the minimum background.

(*e*) At a temperature of 55°C, we observe the recombination radiation of part (*d*) above with a spectrometer whose response is symmetrical, with half-power points separated by a photon energy of 0.120 ev. Discuss the magnitude of the possible errors committed by interpreting the peak of the spectrometer response to the emitted radiation as being the peak of the actual spectral line. This question is also relevant to the experiment performed in the film referred to on p. 59.

P3.3 This problem is concerned with analyzing the random fluxes of particles in a dilute "particle gas," like that of conduction electrons (or holes) in a semiconductor. You can use this analysis immediately to improve your understanding of the mechanism of diffusion, and later it will be helpful in connection with your study of the current flow in *pn* junctions.

Consider a differential element of area dA located at the origin of coordinates, and oriented normal to the x-axis. You wish to evaluate the random flux dF_{LR} of particles from left to right across dA, which means the number of particles crossing dA in unit time from points $x < 0$ to the left of the plane $x = 0$. To this end you must find all particles which have the following properties: they make their last collision in a volume dV located at spherical coordinates (r,θ,ϕ) from the origin at dA; they are also headed for dA; and they will make no further collisions between dV and dA. Then you must add up the contributions from all such volumes dV for which $x = r\cos\theta \leq 0$.

(*a*) Treat a group of $n(v)\,dv$ particles having speeds within dv of v. If $l(v)$ is the *mean free path* of such particles, show that $\dfrac{vn(v)}{l(v)}\,dv\,dV$ is the number of collisions per second made by this group within dV. This is the number of "free paths" for this speed group that *start* in dV per unit time.

(*b*) If the particles leave the collisions with a uniform spatial distribution of velocities, use the solid angle subtended by dA at dV to compute the fraction of particles headed toward dA.

(*c*) Show from the definition of the "mean free path" $l(v)$ that, in general, the fraction of particles which do *not* make a collision in a distance r is $e^{-r/l(v)}$.

(*d*) Thus show that the number of particles that have collisions in dV in unit time, *and* that cross dA before making another collision, can be written

$$d^3N(v) = \left(\frac{1}{4\pi}\right)\frac{n(v)v}{l(v)}\,dA\,\frac{\cos\theta}{r^2}\,dv\,dVe^{-r/l(v)}$$

$$= \frac{dA}{4\pi}\,\frac{n(v)v}{l(v)}\,e^{-r/l(v)}dv\,dr\,\cos\theta\,\sin\theta\,d\theta\,d\phi$$

(*e*) Carrying out all the integrations, show that

$$dF_{LR} = \tfrac{1}{4}n <v> dA$$

where

$$n = \int_0^\infty n(v)\,dv \quad \text{and} \quad <v> = \frac{1}{n}\int_0^\infty vn(v)\,dv$$

(*f*) Repeat the calculation for a case in which n varies linearly with position

$$n = n_o + \left(\frac{dn}{dx}\right)x = n_o + \left(\frac{dn}{dx}\right)r\cos\theta$$

and dn/dx is constant. Evaluate $dF_{LR} - dF_{RL}$ for this case and show that

$$F_{\text{diffusion}} = \frac{dF_{LR}}{dA} - \frac{dF_{RL}}{dA} = -\tfrac{1}{3}<v^2\tau(v)>\frac{dn}{dx} \equiv -D\frac{dn}{dx}$$

where $\tau(v) = \dfrac{l(v)}{v}$ is the mean free time of the group, and

$$<v^2\tau(v)> = \frac{1}{n}\int_0^\infty v^2\tau(v)n(v)\,dv.$$

(*g*) Use the value of D from (*f*), and the mobility from Ch. 1, to evaluate D/μ. If $\tau(v)$ is *independent* of v, and the *equilibrium* distribution-density for the particles is of the Boltzmann form, evaluate the necessary integrals to obtain the Einstein relation. Note that $n(v)$ above is for *equilibrium*, but $n(E)$ in Ch. 1 is for *nonequilibrium* conditions.

P3.4 Assuming that the effective masses of holes and free electrons have the values suggested in Table 1.0, and that the data of Table 1.1 apply, compare for Ge and Si the predictions of Eq. 3.56 with the results given in Eqs. 3.57. Comment on the reasons for differences.

P3.5 Work out the number of holes per cm³, the position of the Fermi level (use Eqs. 3.51 and 3.52)†, the percentage ionization of the impurities (use Eq. 3.29)†, and the number of free electrons per cm³ in a Ge sample at 300°K to which only Ga has been added, in the concentration 10^{15} atoms/cm³.

P3.6† Is it true that, in material containing donors only, the Fermi level is always nearer to the conduction band than to the valence band, and the reverse is true for material containing acceptors only? Explain why.

P3.7 Compare the concentrations of holes and free electrons, and the position of the Fermi level,† in intrinsic Ge, with the same quantities in Ge containing exactly 10^{16} donors/cm³ *and* 10^{16} acceptors/cm³, all at 300°K. Repeat for 1.1×10^{15} acceptors/cm³ *and* 10^{14} donors/cm³. Compare with P3.5.

P3.8 For Ge with 10^{16} arsenic atoms per cm³, find approximately the temperature at which $p_o = 0.1n_o$. Repeat for Si.

† These parts require more than the minimum background.

4

Nonequilibrium Transport of Charge Carriers

4.0 INTRODUCTION

In Chapter 1 we introduced the important ideas underlying the dynamics of the motion of holes and conduction electrons in semiconductors—namely, *drift* (in response to an electric field), *diffusion* (the natural result of random thermal motion when the concentration is not uniform in space), and *recombination* (the effort to return to the equilibrium concentrations of both types of carriers). In fact, all of these processes represent the response of the semiconductor when it is disturbed from the equilibrium state by external means. We now wish to consider such processes quantitatively.

Let several types of nonequilibrium forces act together, but suppose each one is small enough so that, acting alone, it would produce only a small departure of the system from equilibrium. Then the resulting effect of all of them may be considered to be a simple superposition of the individual effects. That is, the effects do not mutually interact. Moreover, the individual results are linearly dependent on their causes. Thus we pointed out in Chapter 3, Eq. 3.65, that the current densities of holes and conduction electrons, produced by the simultaneous action of a small electric field and a small concentration gradient are given by the relations

$$\mathbf{J}_h = q\mu_h p\mathbf{E} - qD_h\nabla p \tag{4.1a}$$

$$\mathbf{J}_e = q\mu_e n\mathbf{E} + qD_e\nabla n \tag{4.1b}$$

$$\mathbf{J} = \mathbf{J}_e + \mathbf{J}_h \tag{4.1c}$$

We have seen that under equilibrium conditions in semiconductors with nonuniform impurity distributions, the electric field that exists must be just such as to cancel out the tendency toward diffusion for each type of carrier separately, making both $\mathbf{J}_h$ and $\mathbf{J}_e$ equal to zero. In nonequilibrium situations, which may result from applied voltages or currents or incident light, for example, the individual carrier currents are not both zero in general, and we do not know all the carrier concentrations and the electric field. There are, in fact, two additional important relations needed to solve a flow problem. These are: *the equation of continuity* for each type of carrier, which relates the net time-rate of increase of the number of carriers of a given kind in a region, to the excess current of those carriers entering that region over that leaving it; and *Gauss's law*, which relates the electric field to the charge density (which is, in turn, related to the carrier concentrations). Let us take up these relations.

4.1 THE EQUATIONS OF CONTINUITY

Consider a region of semiconductor occupying volume V and bounded by closed surface S, as shown in Fig. 4.1.† Focus attention on one type of carrier, in this case holes. The number of holes within V is given by the volume integral

$$\int_V p\,dV = \text{number of holes in } V \tag{4.2}$$

The electric current density of holes we have denoted by $\mathbf{J}_h$ amperes per m^2, and, because holes have positive charge q, the

† The treatment here is in three dimensions, and therefore employs quite freely the vector analysis notation. A one-dimensional review of the continuity relationship appears in Sec. 4.3.1. It may be substituted here as the basis of an alternate one-dimensional treatment for all of Secs. 4.1 and 4.2.

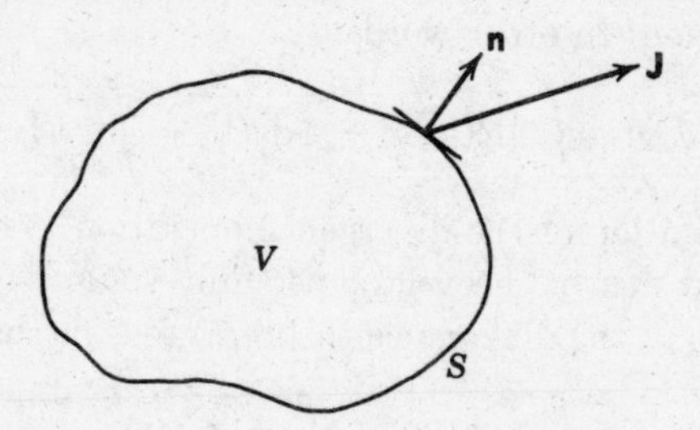

Fig. 4.1. For the development of the continuity equation.

flux density of holes is $\mathbf{J}_h/q$ holes per m² per sec. Therefore, the total flux of holes (number per unit time) *leaving* the region V via its bounding closed surface S is

$$\int_S \frac{1}{q} \mathbf{J}_h \cdot \mathbf{n} \, ds = \text{number of holes leaving } V \text{ per unit time} \tag{4.3}$$

where $\mathbf{n}$ is the *outward* normal to S (Fig. 4.1).

But holes are not lost from V only because they leave; some hole-electron pairs are generated, and some holes recombine with conduction electrons inside V. If the number of recombinations per unit volume, per unit time, is a function $R(n, p)$ of the two carrier concentrations, and if the number of pair generations is G per unit volume, per unit time, the net excess of recombinations over generations taking place within volume V per unit time is

$$\int_V [R(n, p) - G] \, dV = \textit{net} \text{ number of recombinations in } V \text{ per unit time} \tag{4.4}$$

Now the number of holes that disappear from the fixed region V per unit time is

$$-\frac{d}{dt}\int_V p \, dV = -\int_V \left(\frac{\partial p}{\partial t}\right) dV \tag{4.5}$$

That is, the total number disappearing from V per unit time is the sum of all those disappearing in each elemental volume of V per unit time. And this total number must be accounted for by the holes that leave V via its boundary S and those that leave it by recombination. In other words,

$$-\int_V \left(\frac{\partial p}{\partial t}\right) dV = \int_V [R(n, p) - G]\, dV + \int_S \frac{1}{q} \mathbf{J}_h \cdot \mathbf{n}\, ds \qquad (4.6)$$

But the definition of the divergence operation ($\nabla \cdot$) on a vector as the outward flux of the vector per unit volume (or the vector relation known as the divergence theorem) tells us that

$$\int_S \mathbf{J}_h \cdot \mathbf{n}\, ds = \int_V \nabla \cdot \mathbf{J}_h\, dV \qquad (4.7)$$

as long as the derivatives involved in $\nabla \cdot$ are continuous within V. Accordingly, Eq. 4.6 becomes

$$\int_V \left(\frac{\partial p}{\partial t} + R(n,p) - G + \frac{1}{q} \nabla \cdot \mathbf{J}_h\right) dV = 0 \qquad (4.8)$$

The important point now is that Eq. 4.8 must hold for *every* volume V of the semiconductor, however small (provided the necessary derivatives are continuous in V of course). The only way the integral involved can vanish for *every* such choice of V is to require the integrand to vanish. Therefore, the equation of continuity becomes

$$\frac{1}{q} \nabla \cdot \mathbf{J}_h + R(n,p) - G = -\frac{\partial p}{\partial t} \qquad (4.9)$$

We have tacitly been discussing Eqs. 4.6 to 4.9 as if the only process which adds holes to or removes them from circulation within V is that of pair generation or recombination with conduction electrons. This is indeed the usual case. But it does leave out the possibility that separate mobile carriers may be "trapped" temporarily at irregularities or impurities. Such transient "trapping" and "releasing" of the *separate* free carriers is in fact possible, and does occur in some materials at some temperatures. Moreover, our Eq. 4.9 can include such cases if G and R include the emptying and filling rates of the hole traps. But the effect is not important enough in the usual electronic devices for us here to add the complication of keeping track of the transient state of occupancy of such traps, as we should have to do, in addition to following the transient concentrations of holes and conduction electrons.

Thus, on the basis of the foregoing comments, we shall only consider the situation in which every hole that appears or recombines brings or takes with it a conduction electron; so $R(n,p)$ and G are also the number of such electrons recombining or being generated per unit volume, per unit time (P4.1).

Noting that, on account of the negative charge $-q$ of the electron, the *flux* density of electrons is related to the *electrical* current density $\mathbf{J}_e$ by the relationship

$$\mathbf{J}_{\text{particle}, e} = -\frac{1}{q}\mathbf{J}_e$$

we find by methods analogous to those we used for holes the following continuity equation for conduction electrons

$$-\frac{1}{q}\nabla\cdot\mathbf{J}_e + R(n,p) - G = -\frac{\partial n}{\partial t} \qquad (4.10)$$

In general, the generation of pairs may occur both thermally and by other external means, such as light. Thus we can write

$$G = G_{th} + g \qquad (4.11a)$$

where G_{th} is the thermal part and g the balance. As we pointed out in Sec. 1.6.2, there are a good many situations in which the excess of recombination over *thermal* generation takes place with a simple lifetime. When the lifetime idea applies, recombination is expressed in terms of the *minority* carrier, as follows,

$$R(n,p) - G_{th} = \frac{p - p_o}{\tau_h} = \frac{p'}{\tau_h}; \qquad \text{in } n\text{-type material} \qquad (4.11b)$$

and

$$R(n,p) - G_{th} = \frac{n - n_o}{\tau_e} = \frac{n'}{\tau_e}; \qquad \text{in } p\text{-type material} \qquad (4.11c)$$

in which p' and n' are the *excess* concentrations, above the equilibrium values, of holes and conduction electrons, respectively:

$$p' \equiv p - p_o; \qquad n' \equiv n - n_o \qquad (4.12)$$

On the basis of a simple lifetime description of recombination, then, our two continuity equations, 4.9 and 4.10, as usually written for n-type material become

$$\frac{1}{q}\nabla\cdot\mathbf{J}_h + \frac{p'}{\tau_h} = -\frac{\partial p'}{\partial t} + g \tag{4.13a}$$

$$-\frac{1}{q}\nabla\cdot\mathbf{J}_e + \frac{p'}{\tau_h} = -\frac{\partial n'}{\partial t} + g \tag{4.13b}$$

in which we have used the fact that n_o and p_o are independent of time. A similar set of equations may be written for p-type material (P4.2).

4.2 GAUSS'S LAW

4.2.1 *General Properties*

We come finally to the relationship between the carrier concentrations and the electric field. This relationship stems from one of Maxwell's equations, namely, Gauss's law. If $\mathbf{D}$ is the electric displacement and ρ the charge density, we know that

$$\nabla\cdot\mathbf{D} = \rho \tag{4.14}$$

Now, in the semiconductor there are several sources of the charge density ρ. There are the *fixed* positive charges of the ionized donors N_d^+ per m³ and negative charges of the ionized acceptors N_a^- per m³; and there are the *mobile* positively charged holes p per m³ and the negatively charged conduction electrons n per m³. We shall treat only the situations in which substantially all donors and acceptors are ionized ($N_d^+ = N_d$; $N_a^- = N_a$), which occurs at normal temperatures in most semiconductor devices. (To do otherwise would involve possible transient changes in the extent of ionization of these impurities, and lead us into a discussion of the "trapping" effects mentioned before.) On this basis, the net positive charge density is given by

$$\rho = q(N_d - N_a + p - n) = q(N + p - n) \tag{4.15}$$

in which we note that N is positive for excess donors and negative for excess acceptors. We have seen that $\rho = 0$ everywhere in homogeneous material at thermal equilibrium, and that $\rho \neq 0$

at thermal equilibrium in *nonuniformly* doped material. We shall shortly have considerable discussion about the extent to which ρ is nonzero in uniformly doped material under nonequilibrium conditions, and the extent to which transient changes in ρ are important in any case.

We are usually interested in the electric field **E**, rather than **D**, and for most cases of interest the relationship between the two involves simply the dielectric permittivity ϵ (See Table 1.0).

$$\mathbf{D} = \epsilon \mathbf{E} \tag{4.16}$$

Therefore, in view of Eqs. 4.14, 4.15, and 4.16, we have

$$\nabla \cdot \mathbf{E} = \frac{\rho}{\epsilon} = \frac{q}{\epsilon}(N + p - n) \tag{4.17}$$

There is a feature of Eq. 4.17 that requires special attention. We have a differential equation relating the *divergence* of **E** at a point in space to the value of the charge density at that same point. If the charge density is zero somewhere, **E** has no divergence there—but this does *not* mean $\mathbf{E} = 0$ there! For example, any electrostatic fields in a charge-free vacuum have $\rho = 0$ and $\nabla \cdot \mathbf{E} = 0$, but not $\mathbf{E} = 0$. A good example is the field between parallel plates of an air capacitor. Everywhere *between* the plates there is no charge, so $\nabla \cdot \mathbf{E} = \rho = 0$. Yet $\mathbf{E} \neq 0$. Still, we know very well that the electric field at any point, being the force produced by *other charges* on an imaginary "test" charge at that point, must come from charges *somewhere*; but of course these charges need not be at the place where **E** appears.

Take a very simple example of a uniform semiconductor bar with perfectly conducting end contacts (Fig. 4.2*a*). The steady current density within the bar has an associated electric field **E** related to it by Ohm's law. But this field, like the current density with it, has no divergence anywhere inside the semiconductor. The field is entirely produced by charges on the surfaces of the end contacts, outside (or at least not inside) the semiconductor region under consideration.

It is possible, of course, for a field to be produced both by charges within the region of interest and by those outside it. Such would be the case, for instance, if the bar of Fig. 4.2*a* were of nonuniform

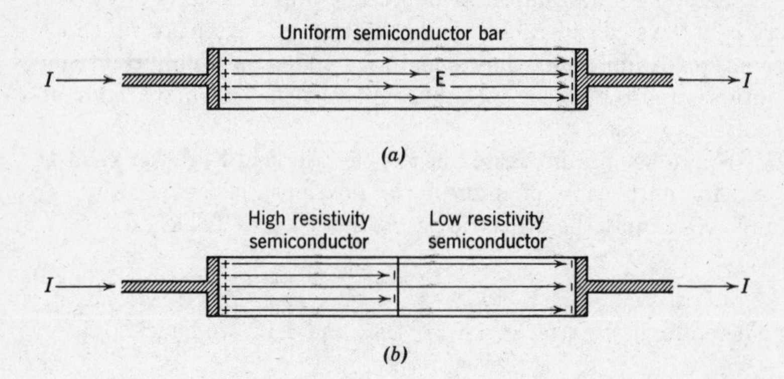

Fig. 4.2. (*a*) Illustrating an electric field, in a uniform neutral semiconductor sample, produced by charges outside the semiconductor. (*b*) Illustrating an electric field produced by charges both inside and outside a nonuniform semiconductor sample.

resistivity, with the left half all of uniformly higher resistivity than the right half (Fig. 4.2*b*). The field on the left would be larger than that on the right, with a negative charge accumulated at the joint to account for the excess of field entering from the left over that leaving to the right—as predicted by Gauss's law. It is not a hard electrostatic problem to show that the three sheets of charge indicated in Fig. 4.2*b*, treated as a one-dimensional problem in an unbounded space having dielectric constant ϵ, will account properly for the zero field at and beyond both perfectly conducting contact interfaces, and for the correct field values in both halves of the bar (P4.3).

Another way of saying more formally that the field in a finite region of space may be produced both by charges in that region and by charges outside it, or on its boundary, is to point out that we can always add any field solution of the equation $\nabla \cdot \mathbf{E}_1 = 0$ to a solution of the equation $\nabla \cdot \mathbf{E}_2 = \rho/\epsilon$, and the sum $\mathbf{E}_3 = \mathbf{E}_1 + \mathbf{E}_2$ will still satisfy the same Gauss's law $\nabla \cdot \mathbf{E}_3 = \rho/\epsilon$. Here $\mathbf{E}_2$ may be thought of as the field produced by charges *inside* the region, and $\mathbf{E}_1$ as the field produced by charges *outside* the region (P4.4).

It should be clear that we are treating the electric field problem in the semiconductor as an (essentially) electrostatic one. Even though we shall sometimes be dealing with time-dependent cases, we shall continue to restrict our treatment to one which omits the effect of the time-varying magnetic field on the solution. This puts an upper frequency limit on the validity of our treatment, which amounts to neglecting "skin effect" in the semiconductor.

4.2.2 *Some Remarks on Electrical Neutrality*

It is worth digressing a little from our direct consideration of nonequilibrium phenomena to point out some features of the departures from electrical neutrality that arise *under thermal equilibrium conditions* in an *inhomogeneously* doped semiconductor containing fully ionized impurities. With Gauss's law, we now have in fact enough relationships to solve such problems.

In particular, if we introduce the potential ψ for the electric field

$$\mathbf{E} = -\nabla\psi \tag{4.18}$$

Eq. 4.17 becomes

$$\nabla^2\psi = -\frac{q}{\epsilon}(N + p - n) \tag{4.19}$$

Then making use of the following Boltzmann-factor relationships from Chapter 3

$$p_o = n_i e^{-q\psi/kT} \tag{3.61}$$

$$n_o = n_i e^{+q\psi/kT} \tag{3.62}$$

in which the arbitrary reference condition $\psi = 0$ has been set at the location where $p_o = n_o = n_i$, we find

$$\nabla^2\psi = \frac{2qn_i}{\epsilon}\left[\sinh\,(q\psi/kT) - \left(\frac{N}{2n_i}\right)\right] \tag{4.20}$$

This equation is, at least in principle, soluble for ψ if $N \equiv N_d - N_a$ is known as a function of position. In practice, a few specific problems have been worked out analytically (P4.5 gives an illustrative example of an approximate method) and other cases have been studied numerically (see PEM, Appendix A).

There is one parameter of importance, however, that we must discuss in connection with the departures from electrical neutrality implied by these equations. If we re-define the electrostatic potential ψ in units of the "thermal voltage" kT/q (which is just kT expressed in electron volts),

$$u \equiv \psi/(kT/q) = q\psi/kT \tag{4.21a}$$

and let

$$\sinh u_o \equiv \frac{N}{2n_i} \tag{4.21b}$$

we have for Eq. 4.20

$$\nabla^2 u = \frac{2q^2 n_i}{kT\epsilon} \quad [\sinh u - \sinh u_o]$$

$$= \frac{4q^2 n_i}{kT\epsilon} \cosh\left(\frac{u + u_o}{2}\right) \sinh\left(\frac{u - u_o}{2}\right) \tag{4.22}$$

Notice that in *uniform* material, for which N is independent of position, Eq. 4.22 is consistent with the fact that u is equal to u_o, with u_o a constant given by Eq. 4.21*b* (check this with Eq. 3.63). This condition results in no unbalanced charge, and no variation of potential, so that both sides of Eq. 4.22 are zero.

But consider a one-dimensional situation like that shown in Fig. 3.3. Near the right-hand end, the material becomes uniformly p-type and $u \rightarrow u_o < 0$ (N is *negative* for p-type material). Take a region where $|u - u_o| \ll 1$, somewhere in the uniform portion. There

$$\sinh\left(\frac{u - u_o}{2}\right) \cong \left(\frac{u - u_o}{2}\right) > 0$$

and

$$\cosh\left(\frac{u + u_o}{2}\right) \cong \cosh u_o = \sqrt{1 + \sinh^2 u_o}$$

$$\cong \frac{N_a}{2n_i} > 0$$

the last approximation holding if the material is reasonably extrinsic ($N_a \gg n_i$). Then Eq. 4.22 may be written

$$\frac{d^2(u - u_o)}{dx^2} = \frac{q^2 N_a}{kT\epsilon} (u - u_o) \tag{4.23}$$

because u_o is independent of x. The solution, which vanishes properly as $x \to +\infty$, is of the form

$$(u - u_o) \sim e^{-x/L_D} \tag{4.24}$$

where

$$L_D \equiv \sqrt{\frac{kT\epsilon}{q^2 N_a}} \tag{4.25}$$

This parameter L_D, called the *Debye length* (or sometimes the *extrinsic* Debye length, because it applies to extrinsic material), is therefore seen to measure the distances over which small variations of potential, measured in thermal volts, smooth out themselves in uniform material.

Actually, because the right-hand side of Eq. 4.22 involves a sinh $(u - u_o)$, it is almost exponential in $u - u_o$ for large departures from neutrality. Therefore, the left side, measuring space-rates of change of the slope of the potential, gets *very* large for large departures from neutrality. For this reason, any *initially large* departures of u from u_o iron themselves out in much *shorter* distances than L_D, and the greatest distances are subsequently consumed by the exponentially vanishing *small* departures. *Therefore, we do not find significant departures from electrical neutrality over distances greater than about* $4L_D$ *to* $5L_D$ *in uniformly doped extrinsic material at thermal equilibrium.*

It is instructive for our later interpretations to rewrite the Debye length in a different way, as follows:

$$\begin{aligned} L_D &= \sqrt{\left(\frac{kT}{q}\right)\left(\frac{\epsilon}{qN_a}\right)} \equiv \sqrt{\left(\frac{kT\mu_h}{q}\right)\left(\frac{\epsilon}{q\mu_h N_a}\right)} \\ &\approx \sqrt{D_h\left(\frac{\epsilon}{\sigma}\right)} = \sqrt{D_h \tau_d} \end{aligned} \tag{4.26}$$

in which $\tau_d \equiv \epsilon/\sigma$ is called the *dielectric relaxation time* (see P1.18). We shall soon find that the time τ_d (usually of the order of 10^{-12} sec), as well as the distance L_D (usually of the order of 10^{-5} cm), are important dimensions associated with departures from electrical neutrality in a semiconductor under *all conditions*, not just for thermal equilibrium.

We shall not dwell any longer on the thermal equilibrium condition here, because we wish to move ahead into the important problems of non-equilibrium flow. These we shall consider only for *homogeneous* material, leaving the further study of inhomogeneous material to be taken up in connection with diodes and transistors in another volume (PEM).

4.3 SOLUTION OF THE FLOW EQUATIONS

4.3.1 *Nature of the Simplifications*

For a discussion of nonequilibrium flow, we have found three vector equations (Eqs. 4.1) and three scalar equations (Eqs. 4.13 and Eq. 4.17) connecting four vector variables ($\mathbf{J}$, $\mathbf{J}_e$, $\mathbf{J}_h$, $\mathbf{E}$) and three scalar variables (p, n, g). Considering that each vector has three scalar components, and each vector equation is really three scalar ones, we therefore have twelve scalar equations and fifteen scalar variables. Thus, in principle, we could solve the whole simultaneous set if we knew, for example, the three scalars or one of the vectors. But as a practical matter, because the differential equations involved are generally nonlinear, an exact simultaneous solution of them is usually impractical, if not impossible.

So we must resort to various approximation methods to get an understanding of what takes place in a flow problem. There is a multitude of special cases we might consider, but there is one broad type which is of overwhelming importance in connection with the most common semiconductor devices. It also forms the important background for developing intuition about other more difficult situations. The cases in question involve three special features:

1. The semiconductor has a *substantial conductivity* (it is not too near to being an insulator).
2. The semiconductor is *extrinsic*; that is, $n_o \gg p_o$ or $p_o \gg n_o$.
3. Any excess carriers introduced have a concentration much less than that of the *majority* carrier; that is p' and n' are much smaller than the *larger* of p_o and n_o. This condition is known as the condition of *small injection*.

The first feature has to do with the extent to which a departure from electrical neutrality (zero charge density) can take place in the uniform regions of the material. After all, for the majority carrier the electrical parameters of such a piece are basically σ

(producing conduction current per unit field) and ϵ (producing displacement current per unit time-rate of change of field). The medium therefore acts like a "leaky" capacitor with an RC time constant $\tau_d = \epsilon/\sigma$ (see also P. 1.18). Even for fairly small values of conductivity like 1 mho/cm, and large relative dielectric constants like 16, this dielectric relaxation time τ_d is extremely short, of the order of 10^{-12} sec, as we pointed out before. *Thus it appears that over other than very short time intervals, it is not possible to maintain any substantial departure from electrical neutrality in the uniform portions of the material.* Any attempt to produce such a departure results in an electric field which, acting through the significant majority-carrier conductivity, produces large conduction currents to neutralize the unbalanced charge rapidly. Correspondingly, for steady-state situations as well as for those involving thermal equilibrium, the *distances* over which departures from neutrality may occur are also extremely short. Specifically, such distances are of the order of a few Debye lengths (Eqs. 4.26 and 4.25, and P4.15, P4.12, and P4.5), which amounts to 10^{-6} to 10^{-4} cm (that is, 100 angstroms to 1 micron).

The other two features listed above together mean that the minority carrier is truly in the minority, under both the equilibrium and nonequilibrium conditions. This not only turns out to make the problem linear, but usually simplifies it considerably (as we have already seen in connection with recombination). Yet we shall find that, even as a small minority, these carriers *can* dominate the electrical behavior of the semiconductor when it is out of equilibrium. Indeed, it is with just such unexpected and interesting dominance of the minority carriers that we shall be primarily (although not exclusively) concerned.

To proceed further with all these points, it is best to look at specific examples. We shall choose one-dimensional cases in order to keep matters as simple as possible. If x is the dimension involved, we have from Eqs. 4.1:

$$J_h = q\mu_h pE - qD_h \frac{\partial p}{\partial x} \tag{4.27a}$$

$$J_e = q\mu_e nE + qD_e \frac{\partial n}{\partial x} \tag{4.27b}$$

$$J = J_e + J_h \tag{4.27c}$$

and from Eqs. 4.9 to 4.11:

$$\frac{1}{q}\frac{\partial J_h}{\partial x} + [R(n,p) - G_{th}] = g - \frac{\partial p}{\partial t} \tag{4.27d}$$

$$-\frac{1}{q}\frac{\partial J_e}{\partial x} + [R(n,p) - G_{th}] = g - \frac{\partial n}{\partial t} \tag{4.27e}$$

and, finally, from Eq. 4.17:

$$\frac{\partial E}{\partial x} = \frac{q}{\epsilon}(N + p - n) \tag{4.27f}$$

In all of these equations the vectors **E** and **J** involve only their x-components E_x, J_{ex}, etc., but to simplify the notation we have here omitted the extra subscript x. On this basis, we can check the interpretation of these one-dimensional results by looking at Eq. 4.27*d* as an example, in connection with Fig. 4.3. The equation of continuity for that figure demands that the net number of particles per unit time *leaving* the volume of small length Δx and unit cross sections at x and $x + \Delta x$, namely,

$$\frac{1}{q}[J_h(x + \Delta x) - J_h(x)]$$

plus the number of excess holes recombining within the volume per unit time

$$[R(n,p) - G_{th}]\,\Delta x$$

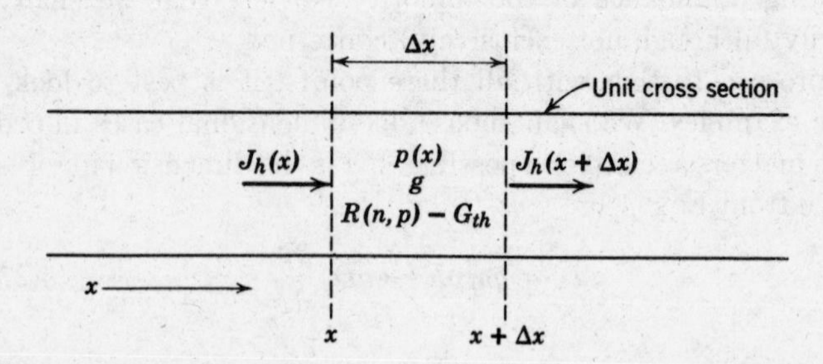

Fig. 4.3. Illustration of the equation of continuity in one space dimension.

plus the increased number of holes in the volume per unit time

$$\frac{\partial}{\partial t}(p\,\Delta x)$$

must equal the number of holes produced per unit time by any excess generation mechanism, like light, in the volume considered:

$$g\,\Delta x$$

Thus, equating all these terms, and dividing by Δx, gives us

$$\frac{1}{q}\left[\frac{J_h(x+\Delta x) - J_h(x)}{\Delta x}\right] + [R(n,p) - G_{th}] + \frac{\partial p}{\partial t} = g$$

keeping in mind that Δx and t are independent variables. In the limit as $\Delta x \to 0$, therefore, we have

$$\frac{1}{q}\frac{\partial J_h}{\partial x} + [R(n,p) - G_{th}] = g - \frac{\partial p}{\partial t}$$

which agrees with Eq. 4.27*d*.

4.3.2 *An Illustrative Steady-State Example*

Consider a very long bar of *uniform* semiconductor, which we take to be extrinsic, n-type. Suppose it is open-circuited, but is uniformly illuminated with a *steady* light over a region of length δ at its center, as shown in Fig. 4.4*a*. Assume the light penetrates uniformly across the cross section and produces hole-electron pairs at a rate g pairs per m^3 per sec. We shall treat the problem as one-dimensional, of unit cross-sectional area. We can see that carrier pairs produced near the middle will tend to diffuse outward, giving rise to hole and conduction-electron diffusion currents in opposite directions. There may also be some electric fields produced, which will cause hole and conduction-electron currents to flow in the same direction. Each of these currents will approach zero at large distances from the light disturbance. Of course, the *sum* of the hole and electron currents at any cross section must be zero, on account of the open-circuited condition of the bar; but the hole and electron currents individually need *not* vanish, because the bar is not in thermal equilibrium.

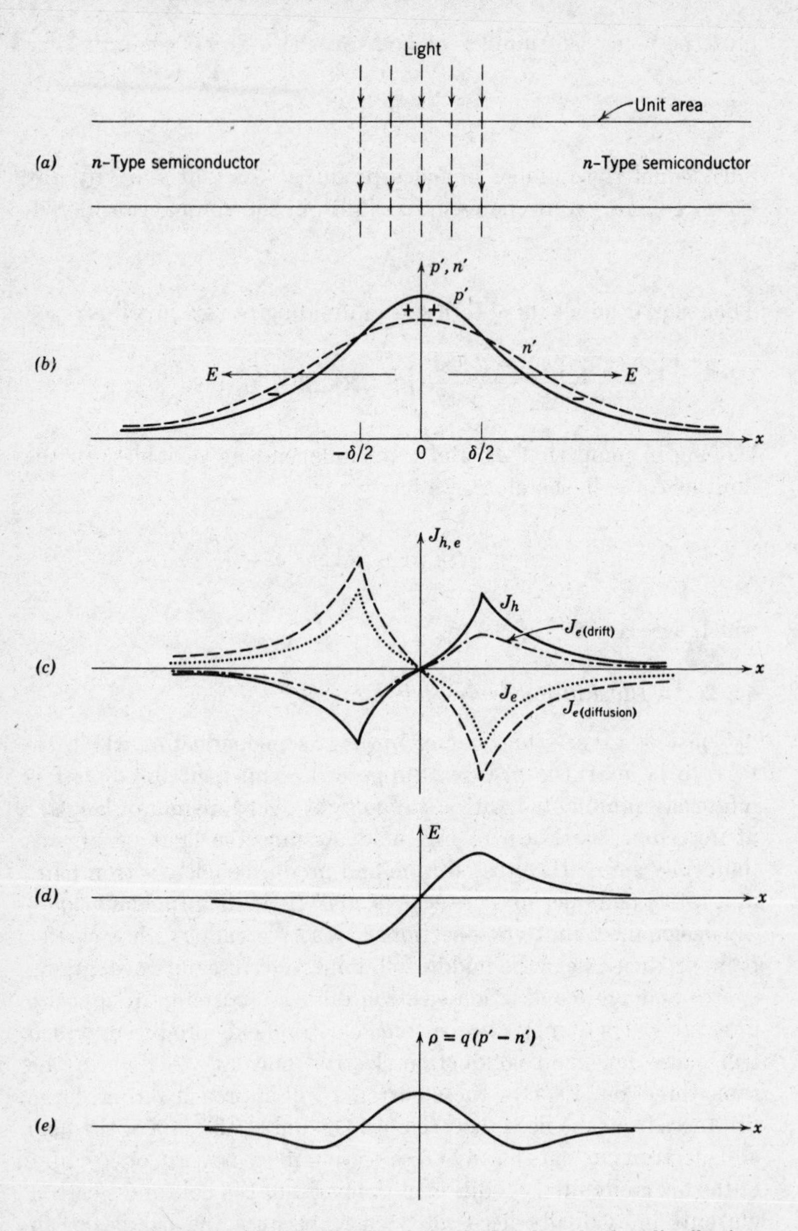

Fig. 4.4. Minority-carrier distribution (p') calculated, and other distributions obtained from qualitative reasoning, for a partially illuminated uniform semiconductor bar.

Inasmuch as the bar and light source are symmetrical about $x = 0$, the distribution of excess carriers will also be symmetrical, and we need consider the solution only for $x > 0$.

For the uniform bar, however, the equilibrium concentrations n_o and $p_o \ll n_o$ are not functions of position, so dn/dx and dp/dx in Eqs. 4.27 may be replaced by dn'/dx and dp'/dx, respectively. Therefore, Eqs. 4.27*a* and *b* become

$$J_h = q\mu_h pE - qD_h \frac{dp'}{dx} \tag{4.28a}$$

$$J_e = q\mu_e nE + qD_e \frac{dn'}{dx} \tag{4.28b}$$

Also, the open-circuited condition of the bar requires that

$$J = J_e + J_h = 0 \tag{4.28c}$$

To go further, we need to assess the relative sizes of the drift and diffusion terms in Eqs. 4.28*a* to *c*. To this end, we recall that the discussion of electrical neutrality in the last section certainly suggests that the steady-state values of n' and p' cannot differ very much at any point if the conductivity of the bar is not too small. The large field that would result from any significant charge unbalance would cause excessive current to flow. Note in this connection that the neutrality of the uniformly doped bar under equilibrium conditions implies that $N + p_o - n_o = 0$, *so the condition of nearly over-all neutrality in the nonequilibrium state is that* n' *and* p' *differ very little at every point.*

But let us be very careful about the sense in which we mean this. What we really wish to assume is that *the variations in the concentrations of excess carriers of both types are very similar with position, so their difference is small compared to either one. That is,*

$$\left|\frac{p' - n'}{p'}\right| \approx \left|\frac{p' - n'}{n'}\right| \ll 1$$

We do *not* wish to imply that $p' - n' = 0$, because this would have a *profound* effect on E, according to the present form of Eq. 4.27:

$$\frac{dE}{dx} = \frac{q}{\epsilon}(p' - n') \tag{4.28d}$$

Such effects in E may be reflected by major errors in the drift component of the current of *majority* carriers (here n) through the term $q\mu_e nE$. In other words, with the very large majority-carrier concentrations involved in these problems, very small electric fields and variations in field can nevertheless produce important majority currents. Neglecting $p' - n'$ on the right side of Eq. 4.28*d* is therefore a much more serious step than neglecting $p' - n'$ compared to either p' or n', which we are perfectly prepared to do. Accordingly, in view of the general similarity of the behavior of both types of carriers, as expressed by our equations, dp'/dx and dn'/dx are also expected to differ by little, in the same kind of sense we employed for p' and n'. So, as long as the diffusion coefficients D_e and D_h are not radically different, the diffusion terms $D_e(dn'/dx)$ and $D_h(dp'/dx)$ of Eqs. 4.28*a* and *b* are of comparable magnitudes.

Now we are presently in a situation where *minority* carriers have *major* effects, because the magnitude of the hole current is comparable to that of the electron current in Eqs. 4.28. But because $n \gg p$, and in the case that μ_e and μ_h are comparable in size, the drift term $q\mu_e nE$ for the majority electrons is much larger than that for the minority holes, $q\mu_h pE$. This is no more than a statement of the fact that, except in cases of unusual mobilities, the minority carrier contributes negligibly to extrinsic conductivity. Thus, for the minority-carrier current J_h to be even comparable in size to the majority-carrier current, the *second* term (the diffusion term) in J_h must dominate the first. Then it becomes a fair approximation to say that

$$J_h \approx -qD_h \frac{\partial p'}{\partial x} \tag{4.29}$$

In fact, we can extend our reasoning more generally and conclude that, *under conditions of comparable mobilities and small injection in uniform extrinsic material, the minority-carrier current will be comparable to the majority-carrier current only if the minority carriers flow mainly by diffusion.* Note that this analysis also implies that the drift and diffusion contributions to majority-carrier current may then be comparable to each other.

Let us agree to look for a solution to our present problem on the basis of Eq. 4.29 for the minority carrier. If we can solve for the minority-carrier current J_h first, J_e will be relatively easy to find from the fact that $J = 0$. We can then check to see if our approximations were valid by using the approximate solutions we obtain to evaluate the size of neglected terms.

To proceed with the solution, we focus attention on the minority carriers, and observe that in the steady state $\partial p/\partial t \equiv 0$. We also take the recombination to be in the form of a simple lifetime τ_h for the excess minority carriers. In the region $x > \delta/2$ where there is no light, $g = 0$. Using Eqs. 4.29 and 4.11*b* in Eq. 4.27*d* therefore yields

$$\frac{d^2p'}{dx^2} - \frac{p'}{D_h\tau_h} = 0; \qquad x > \delta/2 \tag{4.30}$$

The solution which vanishes at $x = +\infty$, rather than becoming infinite there, is the one we need, on the grounds that the effect of the source must vanish far away. It is

$$p' = Be^{-x/L_h}; \qquad x > \delta/2 \tag{4.31}$$

where

$$L_h \equiv \sqrt{D_h\tau_h} \tag{4.32}$$

is called *the diffusion length* for excess holes in the *n*-type material (P4.12 and P4.15), and B is a constant. The corresponding hole current density from Eq. 4.29 is

$$J_h = \frac{qD_h}{L_h} Be^{-x/L_h}; \qquad x > \delta/2 \tag{4.33}$$

Next, we must take account of the region $0 < x < \delta/2$, where light generates carrier pairs at a constant rate $g = g_L$ pairs per m³ per sec. In place of Eq. 4.30 we find for this region

$$\frac{d^2p'}{dx^2} - \frac{p'}{L_h{}^2} = -\frac{g_L}{D_h}; \qquad 0 < x < \delta/2 \tag{4.34}$$

The solution is composed of two parts. First, there is the complementary solution, which is a solution to Eq. 4.34 with the right side zero. This differential equation is similar to Eq. 4.31, but the solution we now need is of the form

$$p' = Ce^{-x/L_h} + De^{+x/L_h}$$

This time the behavior of the solution at $x = \infty$ imposes no constraint on the choice of C and D, because $x = \infty$ is *not* one of the points in the limited region of space on the bar now being considered.

The second part of the solution to Eq. 4.34 is a particular solution

$$p' = g_L \tau_h$$

as we can easily justify by substitution. The meaning of this one is pretty clear, if we note that $g_L \tau_h$ is the total concentration of excess holes produced by the uniform light in one lifetime τ_h. If the entire problem were a uniformly lit bar, the excess concentration $g_L \tau_h$ would be exactly the one for which we would obtain a steady state, because the recombination rate would then be $p'/\tau_h = g_L$, which balances the optical production rate. Of course, the entire problem we have in Fig. 4.4*a* is *not* uniform and that is why we need *both* the particular solution (which does not meet all our boundary conditions) *and* the complementary solution (which, as we shall soon see, has just enough arbitrary constants in it to fit all the required boundary conditions). Thus

$$p' = Ce^{-x/L_h} + De^{+x/L_h} + g_L \tau_h; \qquad 0 \leq x < \delta/2 \qquad (4.35)$$

The hole current within the illuminated region is then given by Eqs. 4.29 and 4.35 as

$$J_h = \frac{qD_h}{L_h}\left[Ce^{-x/L_h} - De^{+x/L_h}\right]; \qquad 0 < x \leq \delta/2 \qquad (4.36)$$

This current must vanish at $x = 0$. If it did not, the symmetry of the problem would demand that it either diverge on the left and right from the plane $x = 0$, or converge from the left and right onto this plane. Neither of these choices is acceptable because, if either were true, the continuity equation (Eq. 4.6) would then require either a finite steady-state net rate of generation or recombination from a *zero volume* at $x = 0$. This, in turn, calls for an infinite value either of g_L or of p' at $x = 0$. Neither is the case. So, by using Eq. 4.36, we find that the condition

$$J_h = 0 \text{ at } x = 0$$

requires

$$C = D \qquad (4.37)$$

and the solution 4.36 becomes

$$J_h = -\frac{2qD_hC}{L_h}\sinh\frac{x}{L_h}; \qquad 0 < x < \delta/2 \qquad (4.38)$$

Similarly, at the plane $x = \delta/2$ the hole current given by Eq. 4.38 and that given by Eq. 4.33 must agree. Again this is required to avoid a discontinuity in J_h over a region of zero volume, in which the generation and recombination rates per unit volume are finite. Therefore,

$$Be^{-\delta/2L_h} = -2C \sinh \frac{\delta}{2L_h}$$

or

$$B = C(1 - e^{\delta/L_h}) \tag{4.39}$$

But at the plane $x = \delta/2$ we must also have the hole concentration given by Eq. 4.31 agree with that given by Eq. 4.35 (actually, it is total hole concentrations $p = p_o + p'$ which must agree on both sides of $x = \delta/2$, but p_o is constant). The reason for this continuity of carrier concentration lies in the need to avoid the infinite diffusion current J_h which would result from a discontinuity in p'; such a current could not be supplied by any of the finite sources in this problem. Therefore, from Eqs. 4.31, 4.39, 4.35, and 4.37 we find at $x = \delta/2$

$$C(1 - e^{\delta/L_h})e^{-\delta/2L_h} = C(e^{\delta/2L_h} + e^{-\delta/2L_h}) + g_L\tau_h$$

or

$$C = -\frac{g_L\tau_h}{2} e^{-\delta/2L_h} \tag{4.40}$$

The entire solution obtained thus far is, therefore, as summarized below.

$$\left.\begin{aligned} &(a)\quad p' = g_L\tau_h\left(\sinh\frac{\delta}{2L_h}\right)e^{-x/L_h} \\ &(b)\quad J_h = qg_LL_h\left(\sinh\frac{\delta}{2L_h}\right)e^{-x/L_h} \end{aligned}\right\} \quad x \geq \delta/2$$

$$\left.\begin{aligned} &(c)\quad p' = g_L\tau_h\left[1 - e^{-\delta/2L_h}\cosh\frac{x}{L_h}\right] \\ &(d)\quad J_h = qg_LL_he^{-\delta/2L_h}\sinh\frac{x}{L_h} \end{aligned}\right\} \quad 0 \leq x < \delta/2 \tag{4.41}$$

The results for p' are shown as solid lines in Fig. 4.4*b*, and those for J_h as solid lines in Fig. 4.4*c*. We can see how the minority holes, which we have assumed flow primarily by diffusion, produce large flow where the concentration gradient is large and small flow where it is small.

Now we know that the total current in the bar must be zero, so the entire majority electron current J_e *must* be equal and opposite to J_h at every point, as shown by the dotted line in Fig. 4.4*c*. In general we have pointed out that the majority carrier will be influenced strongly by both concentration gradients and electric fields. Once the minority-carrier current has been determined in problems like the present one, the majority carrier adjusts itself to do what is necessary to accommodate the other conditions of the problem. It is worthwhile to examine how this comes about in this particular problem, under two kinds of conditions.

First, consider a situation in which $D_e = D_h$ (and therefore also $\mu_e = \mu_h$). Then careful examination of Eqs. 4.27*a* to *c* shows that they are all satisfied if $E = 0$ and $n' = p'$ everywhere (remember that n_o and p_o do not depend on position or time). Similarly, Eqs. 4.27*d* and *e* are also satisfied under the same conditions, because $J_e = -J_h$ along with $n' = p'$. Finally, with $p' = n'$ (and of course $N + p_o - n_o = 0$) there is no unbalanced charge on the right-hand side of Eq. 4.27*f*, which means that $\partial E/\partial x = 0$. The absence of any external source means that no field arises from the end contacts, so $E = 0$ is the acceptable solution. In short, when $D_e = D_h$ the solution to the whole problem implies $p' = n'$ and $E = 0$ everywhere. *Both* types of carrier flow only by diffusion in this problem under these special conditions, and the solution for either excess carrier concentration is exactly like the one we found for the excess minority carrier p'.

But in a typical semiconductor $D_e > D_h$ (in Ge, $D_e \approx 2D_h$, and in Si, $D_e \approx 3D_h$, at 300°K). This means that in our problem the diffusion of the majority carrier is more effective in removing it from the region of optical generation than is the case for the minority carrier. It takes less concentration gradient to get the same magnitude of diffusion current. If, as above, we try to obtain a solution with $E = 0$, we would have to balance the diffusion currents alone of holes and conduction electrons. We would therefore have to make the distribution of conduction electrons wider than that of holes. But because in our problem conduction electrons and holes are generated and recombine only in pairs, we cannot have more *total* excess carriers of one type than the other in the whole bar. Accordingly, the excess-electron distribution curve in Fig. 4.4*b* cannot lie either everywhere above or everywhere below the excess hole curve. It must lie partly above and partly below, in such a way as to have the same area under it. There is no choice but to have a result something like the dashed curve shown in the figure (however, from the discussion thus far, the cross-over point of the two curves need not fall at $x = \pm\delta/2$ as shown).

The foregoing result certainly provides a possible mechanism for balancing just the two diffusion currents when $D_e > D_h$; but, as also shown in Fig. 4.4*b*, there is a resulting pair of dipole charge layers, one on each side of the middle of the bar. Such dipole layers represent a charge density distribution something like that sketched in Fig. 4.4*e*. This, in turn, through Gauss's Law (Eq. 4.28*d*), produces an electric field E indicated schematically by vectors drawn on Fig. 4.4*b* and sketched in form in Fig. 4.4*d*.

The electric field just encountered produces a majority-carrier drift component of current in the *same* direction as the diffusion current of holes originally found. This is indicated by the dash-dot curve in Fig. 4.4*c*. This curve is proportional to the one for E because the electron drift current is given by

$$J_{e(\text{drift})} = q\mu_e nE = q\mu_e(n_o + n')E \approx q\mu_e n_o E \qquad (4.42)$$

an approximation justified by the assumption of small injection. We have already indicated that the drift current of holes $q\mu_h pE = q\mu_h(p_o + p')E$ is negligible compared to its diffusion current, because the total minority-carrier current is as large as that of the majority carrier.

Thus, when $D_e \neq D_h$ in this problem, it develops that the majority-carrier diffusion current must not only oppose and cancel the minority-carrier diffusion current, but it must also cancel the drift component of majority-carrier current. This requires a steeper gradient of n' than appeared to be needed just to balance the ratio (often of order only 2 or 3) between D_e and D_h. Making steeper the dashed curve in Fig. 4.4*b* does two things, however. It gives the additional electron diffusion current, toward the center, that is needed to oppose the effect of the field in our problem; and it reduces the unbalanced charged dipole layer which is, in turn, producing that very field. Clearly, a balance exists for which the electron diffusion current (dashed in Fig. 4.4*c*) will over-oppose its drift component (dash-dot curve in Fig. 4.4*c*) by just enough so the resultant (dotted curve in Fig. 4.4*c*) can just cancel the minority current, as required. Notice that the electron diffusion current needed when $D_e > D_h$ cannot exceed $(D_e/D_h)J_h$ in magnitude, because if it had even this magnitude, n' would equal p' everywhere, and there would be no source of unbalanced charge left to make the field which needs to be opposed!

Nevertheless, because it takes *so little difference* between p' and n' to produce enough field to adjust the majority-carrier current a great deal, we can in fact make a reasonable *estimate* of all the currents and the electric field by assuming that

$$\begin{aligned} &(a) \quad p' \approx n' \\ &(b) \quad \frac{dp'}{dx} \approx \frac{dn'}{dx} \end{aligned} \tag{4.43}$$

and, therefore, that

$$\begin{aligned} J_{e(\text{diffusion})} &= qD_e \frac{dn}{dx} = qD_e \frac{dn'}{dx} \\ &\approx qD_e \frac{dp'}{dx} \approx -\frac{D_e}{D_h} J_h \end{aligned} \tag{4.44}$$

In Fig. 4.5*a* and *b* we have repeated Fig. 4.4*a* and part of 4.4*b*, respectively, and in Fig. 4.5*c* we show the result of Eq. 4.44 as a dashed line, drawn for $D_e/D_h \approx 2$.

As we have said before, this diffusion current of electrons cannot be the entire electron current. There must be a drift term for the electrons which just makes the total current zero:

$$J_{e(\text{drift})} = -(J_{e(\text{diffusion})} + J_h) \approx J_h \left(\frac{De}{D_h} - 1\right) \tag{4.45}$$

The dash-dot line in Fig. 4.5*c* illustrates this component of the electron current which, when added to the diffusion component, we found before produces a total of

$$J_e = J_{e(\text{drift})} + J_{e(\text{diffusion})} = -J_h \tag{4.46}$$

as shown by the dotted line in Fig. 4.5*c*.

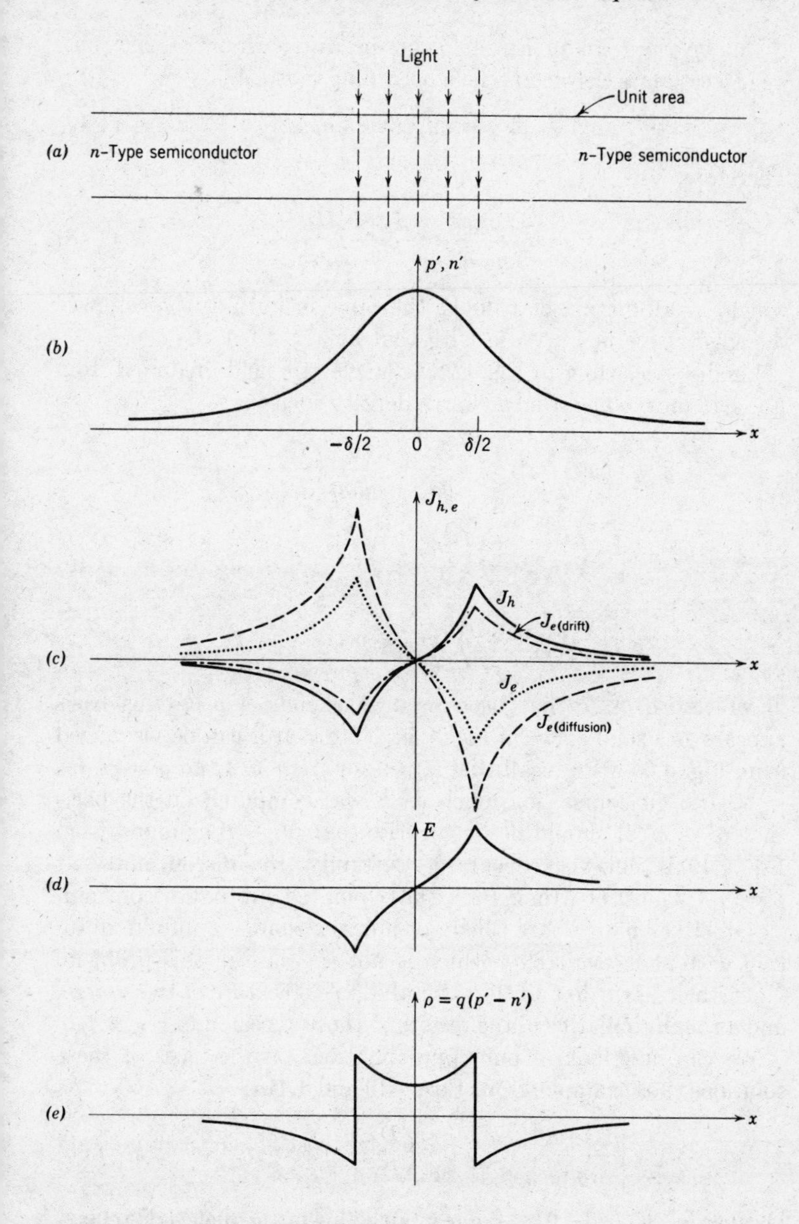

Fig. 4.5. An approximate quantitative solution for the problem of an illuminated semiconductor bar, based on the approximation $n' \approx p'$. Note serious errors in ρ.

But on the basis of Eq. 4.45 for the drift current of electrons, we can estimate the electric field according to Eq. 4.42:

$$J_{e(\text{drift})} = q\mu_e nE \approx q\mu_e n_o E \tag{4.47}$$

Therefore,

$$E \approx \frac{J_{e(\text{drift})}}{q\mu_e n_o} \approx \frac{J_h\,(D_e/D_h - 1)}{q\mu_e n_o} \tag{4.48}$$

which, as a function of x, looks the same in form as J_h (compare the solid curve in Fig. 4.5*d* with that in Fig. 4.5*c*).

Finally, according to Eq. 4.28*d*, the electric field predicted from Eq. 4.48 must come from a charge density such that

$$\begin{aligned} \rho = q(p' - n') &\approx \epsilon\frac{dE}{dx} = \frac{\epsilon}{q\mu_e n_o}\left(\frac{D_e}{D_h} - 1\right)\frac{dJ_h}{dx} \\ &\approx \frac{\epsilon}{\sigma}q\left(\frac{D_e}{D_h} - 1\right)\left(g - \frac{p'}{\tau_h}\right) \\ &\approx q\frac{\epsilon}{\sigma\tau_h}\left(\frac{D_e}{D_h} - 1\right)(g\tau_h - p') \end{aligned} \tag{4.49}$$

in which Eq. 4.27*d* has been used. A sketch of ρ on this basis appears as a solid curve in Fig. 4.5*e*. Note that it can be visualized from Fig. 4.5*b* if we recall that $g = 0$ for $|x| > \delta/2$ and $g = g_L$ for $|x| < \delta/2$. Of course, inasmuch as E was computed on the basis that $p' = n'$, it should be no surprise that $p' - n'$ as found from Eq. 4.49 is not very accurate. Certainly, the discontinuity at $x = \pm\delta/2$ must be wrong, for we have pointed out that discontinuities in either p' or n' are ruled out unless a source of infinite diffusion current is available—which is not the case in this problem. Nevertheless, we do find the general dipole structure of the charge, and a rough similarity of the results to those expected in Fig. 4.4.

We can now look at our approximations on the basis of these solutions. For example, from Eqs. 4.49 and 4.41*a*:

$$\left|\frac{p' - n'}{p'}\right|_{x \geq \delta/2} \approx \frac{\tau_d}{\tau_h}\left(\frac{D_e}{D_h} - 1\right) \approx 10^{-6} \tag{4.50}$$

because $\tau_d = \epsilon/\sigma \approx 10^{-12}$ sec is a fair value of the dielectric relaxation, and $\tau_h \approx 10^{-6}$ sec is a representative minority-carrier lifetime.

The approximation of "nearly neutral" or "quasi-neutral" behavior is therefore not marginal, but is extremely well-justified in most cases. It is usually satisfied by a large margin.

The other approximation we have been making concerns the ratio of drift to diffusion components of minority-carrier current, which we have assumed is very small. Again, for $x \geq \delta/2$ as an example, Eqs. 4.41*a* and *b* with Eq. 4.48 indicate that

$$\left|\frac{q\mu_h pE}{qD_h(dp/dx)}\right| = \frac{q\mu_h(p_o + p')J_h\,(D_e/D_h - 1)}{J_h q\mu_e n_o}$$
$$= \left(\frac{p_o + p'}{n_o}\right)\left(1 - \frac{D_h}{D_e}\right) \ll 1 \tag{4.51}$$

because of the comparable mobility (D_e and D_h not very dissimilar in magnitude), small-injection ($p' \ll n_o$), and extrinsic ($p_o \ll n_o$) conditions on the problem.

The conclusion we wish to emphasize again in this connection *is that, for small injection in homogeneous extrinsic material, the minority-carrier current must flow primarily by diffusion if its magnitude is to be comparable to the majority-carrier current (provided the mobility values for both carriers are comparable)*. Then the ready supply of majority carriers effectively "shields" the minority ones from producing any significant space charge. The small fields that are generated by *slight* departures from electrical neutrality serve to adjust the majority-carrier current to the general conditions of the problem, without producing significant effects on the minority carriers. Thus an approximate calculation of J_h, J_e, E, p', and n' on the "nearly neutral" or "quasi-neutral" basis $p' \approx n'$ (but *not* using Gauss's law with $p' - n' = 0$) will be quite satisfactory. Of course, the small $p' - n'$ will usually not be very accurately determined from E found in this way.

If we are disturbed by the erroneous cusps in $J_{e(\text{drift})}$ and in E, and the resulting false discontinuity they generate in the very small $p' - n'$ in our approximate solution (Fig. 4.5, as compared to Fig. 4.4), it is possible to improve the solution considerably—but with a great deal more work. This is done in the next section for those who are especially interested in fine details. A sketch of the results is however presented for immediate comparison in Fig. 4.6, where it is clear that all the difficulties are removed.

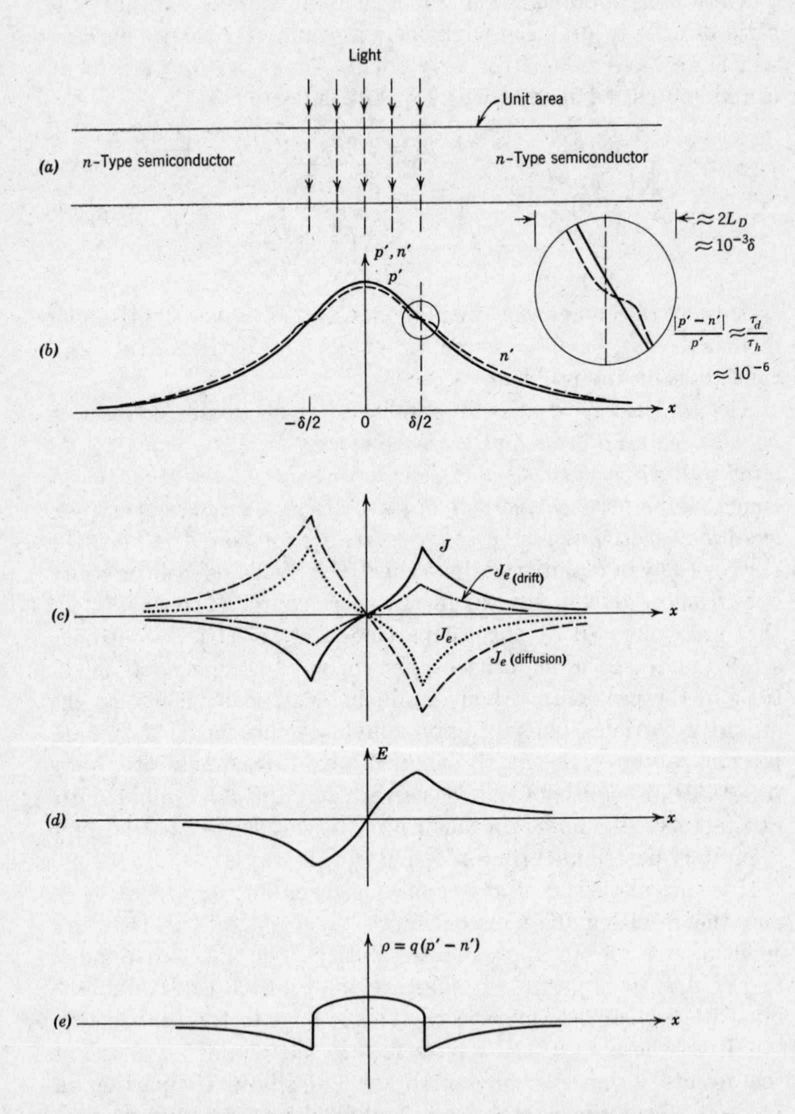

Fig. 4.6. A nearly exact solution for an illuminated semiconductor bar. Compare with Figs. 4.4 and 4.5.

4.3.3 *A More Accurate Solution*

To make a more accurate solution of our problem (see Fig. 4.6), instead of using $p' \approx n'$ to evaluate $J_{e(\text{diffusion})}$ as we did in Eq. 4.44, we can write directly the current balance $J_e = -J_h$ for open-circuit conditions in the form

$$q\mu_e n_o E + qD_e \frac{dn'}{dx} = -J_h$$

(where we neglected $n'E$ again). Then we can differentiate it, and on the right-hand side use the continuity relation for holes. Next, we use Gauss's law (Eq. 4.28*d*) for dE/dx, with the result:

$$\frac{\sigma}{\epsilon}(p' - n') + D_e \frac{d^2n'}{dx^2} = \frac{p'}{\tau_h} - g$$

Finally, introducing the dielectric relaxation time τ_d, and dividing by D_e, yields

$$\frac{d^2n'}{dx^2} + \frac{p' - n'}{L_D^2} = \frac{p'}{bL_h^2} - \frac{g}{D_e} \tag{4.52a}$$

where $b \equiv D_e/D_h = \mu_e/\mu_h$ is the mobility ratio, and $L_D = \sqrt{D_e\tau_d}$ is the Debye length (for the majority electrons). If we regard p' as found in Eq. 4.41 as correct (and, of course, g is known), the last linear, second-order differential equation can be solved for n' for the regions $|x| < \delta/2$, where $g = g_L$, and $|x| > \delta/2$, where $g = 0$. Both particular and complementary solutions must be found for *both* regions now, because of the known p' terms in Eq. 4.52*a*. Boundary conditions must be applied to n' at $x = 0$ and $x = \infty$, as we did for p', and n' and dn'/dx must be required to be continuous at $x = \pm\delta/2$ (as was required of p'). Because $L_D \ll L_h\sqrt{b}$ (on account of $D_e \sim D_h$ and $\tau_d \ll \tau_h$), it is clear that very small values of $p' - n'$ will suffice to produce major effects on the value of d^2n'/dx^2. Certainly, the term $(p' - n')/L_D^2$ competes with p'/L_h^2b even when $p' - n' \ll p'$. A full solution is straightforward but a little long. It proceeds as follows.

Collecting all the unknown terms in n' on the left, we get for Eq. 4.52*a*:

$$\frac{d^2n'}{dx^2} - \frac{n'}{L_D^2} = -\left(\frac{1}{L_D^2} - \frac{1}{bL_h^2}\right)p' - \frac{g}{D_e} \tag{4.52b}$$

In the region $x > \delta/2$, we have $g = 0$ and p' given by Eq. 4.41*a*. This p' term yields a particular solution to Eq. 4.52*b* of

$$n'_{\text{part.}} = \left[\frac{1 - (L_D/\sqrt{b}L_h)^2}{1 - (L_D/L_h)^2}\right] p' \equiv \nu p'$$

$$= \nu g_L \tau_h \left(\sinh \frac{\delta}{2L_h}\right) e^{-x/L_h}; \qquad x \geq \delta/2 \tag{4.52c}$$

where we have put

$$\nu \equiv \frac{1 - (L_D/\sqrt{b}L_h)^2}{1 - (L_D/L_h)^2} \tag{4.52d}$$

The complementary solution is

$$n'_{\text{comp.}} = B'e^{-x/L_D}; \qquad x \geq \delta/2 \tag{4.52e}$$

because the term e^{+x/L_D} is unacceptable at ∞.

In the region $|x| \leq \delta/2$, we know that $g = g_L$ and p' is given by Eq. 4.41c. These two driving terms give a particular solution to Eq. 4.52a in this region of

$$n'_{\text{part.}} = g_L\tau_h\left(1 - \nu e^{-\delta/2L_h}\cosh\frac{x}{L_h}\right); \qquad 0 \leq x \leq \delta/2 \tag{4.52f}$$

This expression already has zero slope at $x = 0$, so the complementary solution needs this property *by itself*, because the whole solution needs zero diffusion current of electrons at $x = 0$. Thus,

$$n'_{\text{comp.}} = 2C'\cosh\frac{x}{L_D}; \qquad 0 \leq x \leq \delta/2 \tag{4.52g}$$

Now we need to make dn'/dx continuous at $x = \delta/2$. It turns out that the particular solutions 4.52c and f already have this property, so we need to impose it only on the complementary parts 4.52e and g. The result is a requirement that

$$B' = C'(1 - e^{\delta/L_D}) \tag{4.52h}$$

Finally, we evaluate C' by making n' itself continuous at $x = \delta/2$, employing Eqs. 4.52c, e, h, f, and g. We find after some algebra that

$$\left.\begin{aligned} C' &= \left(\frac{\nu - 1}{2}\right) g_L\tau_h e^{-\delta/2L_D} \\ B' &= (1 - \nu) g_L\tau_h \sinh\frac{\delta}{2L_D} \end{aligned}\right\} \tag{4.52i}$$

so the result for n' is

$$n' = g_L\tau_h\left[1 - \nu e^{-\delta/2L_h}\cosh\frac{x}{L_h} + (\nu - 1)e^{-\delta/2L_D}\cosh\frac{x}{L_D}\right]; \qquad 0 \leq x \leq \delta/2 \tag{4.52j}$$

$$n' = g_L\tau_h\left[\nu\left(\sinh\frac{\delta}{2L_h}\right)e^{-x/L_h} - (\nu - 1)\left(\sinh\frac{\delta}{2L_D}\right)e^{-x/L_D}\right]; \qquad x \geq \delta/2$$

Then the critical item which was poorly determined previously becomes

$$\frac{p' - n'}{g_L \tau_h} = (\nu - 1)\left[e^{-\delta/2L_h} \cosh \frac{x}{L_h} - e^{-\delta/2L_D} \cosh \frac{x}{L_D}\right]; \qquad 0 \le x \le \delta/2 \quad (4.52k)$$

$$\frac{p' - n'}{g_L \tau_h} = (\nu - 1)\left[\left(\sinh \frac{\delta}{2L_D}\right) e^{-x/L_D} - \left(\sinh \frac{\delta}{2L_h}\right) e^{-x/L_h}\right]; \qquad x \ge \delta/2$$

Now, in the common case, $L_D \ll L_h$ and $\sqrt{b} > 1$. Thus $1 \gg \nu - 1 > 0$. If we select a situation where δ is of the order of L_h (so $\delta \gg L_D$), and make sketches on a scale with L_h and δ as significant distances, the behavior of our solution 4.52k is shown by Fig. 4.6. The apparent discontinuities in ρ are now actually exponentials which occupy dimensions of a few L_D, as suggested in the inset of Fig. 4.6b, and the major troubles with the earlier solution have been cured. Incidentally, we can learn here that the Debye length L_D is a measure of the smallest *distance* within which an unbalanced charge can neutralize itself in the conductor under steady-state conditions in time. It is thus the spatial counterpart of the dielectric relaxation time τ_d, just as the diffusion length L_h is the spatial counterpart of the lifetime τ_h. In fact, as we have seen $(L_D/L_h)^2 = \tau_d/\tau_h$.

4.3.4 *An Illustrative Transient Example*

Suppose we return to the long bar of our previous example (Fig. 4.6a) and imagine that the light source is a single short-duration flash, rather than a steady light. But to make the problem as simple as possible, let us use a very intense line source which produces an extremely *large* number g_L of excess carrier pairs per unit volume per unit time, over an extremely small length δ and an extremely short time $T(\ll \tau_h)$. The net result is that when such a flash is applied to a bar previously at equilibrium, it establishes virtually instantaneously at $t = 0$ a plane of excess carrier pairs at $x = 0$ having P_o' excess holes per unit area, where

$$P_o' = \lim_{\substack{g_L \to \infty \\ \delta \to 0 \\ T \to 0}} (g_L T \delta) \qquad (4.53)$$

In other words, such a source is being idealized as an impulse in space (x) *and* time (t), of total area P_o'. Of course, there are also P_o'

excess conduction electrons per unit area produced by the flash. We wish to discover what happens after $t = 0$.

Clearly, for $t > 0$ the excess carriers produced at $x = 0$, $t = 0$ will (*a*) diffuse outward to the left and right symmetrically, and (*b*) recombine with each other, as suggested by the sketches of $p'(x, t)$ given in Fig. 4.7. If there were no recombination, neither the number of excess holes, nor the number of excess conduction electrons in the whole bar could change with time after $t = 0$. In that case, the *areas* under all the curves in Fig. 4.7 would be equal:

$$\int_{-\infty}^{\infty} p'(x,t)\, dx = P_o' \qquad \text{for } t > 0 \text{ and no recombination} \tag{4.54}$$

Otherwise, these areas become smaller with time as the two kinds of excess carriers recombine with each other.

To approach a quantitative solution of this problem, we would like to use the approximation of small injection ($p', n' \ll n_o$). Obviously, this approximation cannot be valid at $x = 0$ and $t = 0$ in our idealized problem. It will take a while for it to become valid

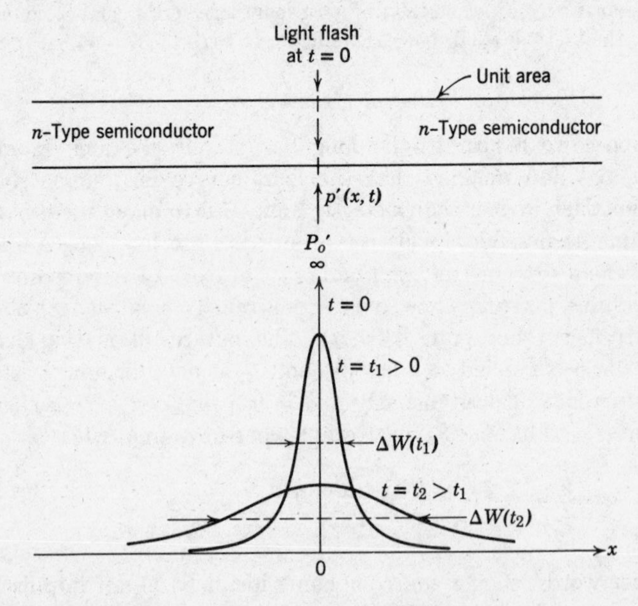

Fig. 4.7. Response of long bar to impulse light source at $x = 0$, $t = 0$.

at $x = 0$ at all in this situation. But the whole problem is really only an idealization of a flash which is short in duration compared to lifetime τ_h, and short in extent compared to the diffusion length L_h. So it makes sense to go ahead with small injection ideas anyway, but only if we are careful to watch out for large injection regions in applying our results to real cases.

We wish to discuss the solution of this problem at all positions and times *except* exactly at the one point $x = 0$, $t = 0$, where the impulse light flash itself occurs. Therefore, $g = 0$ in our problem, and the role of the flash is really just to establish an initial condition of P_o' excess carrier pairs per unit area at $t = 0$, $x = 0$.

Based on the approximations of small injection, quasi-neutrality, and extrinsic n-type material that we used before, the relevant equations for the minority carrier (holes) are Eqs. 4.27d, 4.11b, and 4.29. Combining them, we find

$$\frac{\partial^2 p'}{\partial x^2} - \frac{p'}{L_h^2} = \frac{1}{D_h}\frac{\partial p'}{\partial t}; \qquad -\infty < x < +\infty \text{ and } t > 0 \tag{4.55}$$

It is clear that if $0 < \tau_h < \infty$, new variables $x' = x/L_h$ and $t' = t/\tau_h$ will eliminate all the constants from Eq. 4.55:

$$\frac{\partial^2 p'}{\partial x'^2} - p' = \frac{\partial p'}{\partial t'} \tag{4.56}$$

so p' is really a function of x/L_h (distance in units of diffusion lengths) and t/τ_h (time in units of lifetime). The solution is clearly symmetrical in x or x', because the initial condition has this symmetry and the differential equation is unchanged by substituting $-x$ for x (or $-x'$ for x').

It is actually possible to solve Eq. 4.55 completely for the initial conditions involved here. For $\tau_h = \infty$ it is a *diffusion equation*, common in the theory of heat flow. To make use of this fact we note first that the role of τ_h is quite simple, because physically we have just a condition of diffusion, with the exception of the fact that holes are vanishing at each point at a time rate proportional to the number present. This would seem simply to cause a uniform shrinkage of the space pattern at all points, exponentially in time, without any change of shape. Therefore, we might try to place

$$p'(x,t) = p_\infty'(x,t)e^{-t/\tau_h} \tag{4.57}$$

in Eq. 4.55, with the result that it becomes simply

$$\frac{\partial^2 p_\infty'}{\partial x^2} = \frac{1}{D_h}\frac{\partial p_\infty'}{\partial t} \tag{4.58}$$

a diffusion equation for p_∞' in which recombination is eliminated. A new variable here, $t'' = D_h t$, yields

$$\frac{\partial^2 p_\infty'}{\partial x^2} = \frac{\partial p_\infty'}{\partial t''} \tag{4.59}$$

which shows that p_∞' is actually a function only of x and $D_h t$.

It follows from Eq. 4.57 that the total number of excess holes per unit area in the whole bar at any time, which we shall call $P'(t)$, is just

$$P'(t) = \int_{-\infty}^{\infty} p'(x,t)\,dx = e^{-t/\tau_h}\int_{-\infty}^{\infty} p_\infty'(x,t)\,dx = P_o' e^{-t/\tau_h} \tag{4.60}$$

in contrast to the result 4.54 we had before.

Now we can make use of Eq. 4.59, for which the solution desired has the property of becoming infinite at $t = 0$, $x = 0$, but in such a way that

$$\int_{-\infty}^{\infty} p_\infty'(x,t'')\,dx = P_o', \qquad \text{at } all\ t'' \geq 0. \tag{4.61}$$

This solution is

$$p_\infty'(x,t'') = \frac{P_o'}{2\sqrt{\pi t''}}e^{-x^2/4t''} = \frac{P_o'}{2\sqrt{\pi D_h t}}e^{-x^2/4D_h t} \tag{4.62}$$

We can check this solution in Eq. 4.59 by direct differentiation, and we can actually perform the integration of it required in Eq. 4.61 by using the integrals in Eqs. 3.17 and 3.18.

Based on the result in Eq. 4.62, we can see that at any time $t_1'' > 0$, the "width" of the distribution measured, for example, by the distance $\Delta W(t_1'')$ between the two symmetrically located places where it has one-half of its height at $x = 0$ (see Fig. 4.7), is given by the relation

$$\frac{P_o'}{2\sqrt{\pi t_1''}}\exp - \frac{[\Delta W(t_1'')/2]^2}{4t_1''} = \frac{1}{2}\frac{P_o'}{2\sqrt{\pi t_1''}}$$

or

$$[\Delta W(t_1'')]^2 = (16 \ln 2)t_1'' = 11.09\ D_h t_1 \tag{4.63}$$

In other words, the width at half height is given at any time t by

$$[\Delta W(t)]^2 = 11.09\, D_h t \tag{4.64}$$

This result is characteristic of a diffusion process in one space dimension, where x appears only in the form $(x^2/D_h t)$: *the squared "spread" of the distribution, defined in some appropriate way, is proportional to the product of diffusion coefficient and time measured from the moment of complete condensation.* Alternatively, we may interpret this behavior at the end of time t of a large group of holes that all started at $x = 0$, $t = 0$, in terms of what we would have found by repeatedly starting *one* particle at $x = 0$ and observing where it got to by diffusion at the end of the same time t. In that case, the particle would have ended up at different positions after each trial (positive or negative x of different magnitudes). Any consistent definition of the "squared distance diffused," such as the "average," the "median," or "that not exceeded in some given percentage of the trials," etc., will give a result proportional to $D_h t$. Thus, *according to any reasonable definition, the "squared distance diffused" in time t by a particle in one dimension is proportional to* $D_h t$ (P4.12, P4.15, and P4.21).

Observe that the foregoing results do not depend on the rate of uniform recombination at all. We could have used $p'(x,t)$ instead of $p_\infty'(x,t)$ in obtaining Eq. 4.63, and the fact that all quantities involved are *relative* values at different positions for the *same* time would have dropped out the recombination exponential.

4.3.5 *Modification for Less Extrinsic Material*

In the last section, we treated the problem of Fig. 4.7 on the basis that $n_o \gg p_o$ (this was the "extrinsic" approximation), in addition to the small injection assumption $n', p' \ll n_o$. In case the material is not strongly extrinsic, however, we cannot argue that the minority carrier flows only by diffusion. There is no effective "shielding" of minority by majority carriers, and the minority drift and diffusion terms may be comparable (see Eq. 4.51). We must retain the field and use Eqs. 4.27*a*, and 4.27*d*, with $g \equiv 0$ everywhere except $x = 0$, $t = 0$. Moreover, we must use the recombination law presented for this case in Eq. 1.45. Thus the equation of continuity for holes reads:

$$D_h \frac{\partial^2 p'}{\partial x^2} - \mu_h \frac{\partial (pE)}{\partial x} - \frac{p'}{\tau_1} - \frac{n'}{\tau_2} = \frac{\partial p'}{\partial t}$$

in which τ_1 and τ_2 simply express the constants in Eq. 1.45 conveniently as "lifetimes." Their orders of magnitude are similar to those of τ_e or τ_h. Then

$$\frac{\partial^2 p'}{\partial x^2} - \frac{q}{kT} E \frac{\partial p'}{\partial x} - \frac{q}{kT} p \frac{\partial E}{\partial x} - \frac{p'}{L_1^2} - \frac{n'}{L_2^2} = \frac{1}{D_h} \frac{\partial p'}{\partial t} \tag{4.65}$$

where we have divided by D_h, defined $L_1^2 = D_h\tau_1$ and $L_2^2 = D_h\tau_2$, and used the Einstein relation. Similarly, for the majority carrier

$$-\frac{\partial^2 n'}{\partial x^2} - \frac{q}{kT} E \frac{\partial n'}{\partial x} - \frac{q}{kT} n \frac{\partial E}{\partial x} + \frac{n'}{bL_2^2} + \frac{p'}{bL_1^2} = -\frac{1}{D_e} \frac{\partial n'}{\partial t} \tag{4.66}$$

in which $b \equiv D_e/D_h = \mu_e/\mu_h$.

Now the sensitive term is $\partial E/\partial x$, because it depends on $p' - n'$ through Gauss's law, Eq. 4.28*d*, with a coefficient $1/L_D^2$. This is a situation similar to the one in Eq. 4.52*a*, and here, as there, the term $(p' - n')/L_D^2$ can easily compete with a term like n'/L_2^2, even when $p' - n' \ll n'$, because $L_D \ll L_2$. That is, the small term $p' - n'$ may not be negligible here, because it may be multiplied by a very big number. Thus it would be best to avoid estimating this term, if possible. We can, in fact, eliminate this term from Eqs. 4.65 and 4.66 by division of the first by p, the second by n, and subtraction of the results:

$$\frac{1}{p}\frac{\partial^2 p'}{\partial x^2} + \frac{1}{n}\frac{\partial^2 n'}{\partial x^2} + \frac{q}{kT} E \left(\frac{1}{n}\frac{\partial n'}{\partial x} - \frac{1}{p}\frac{\partial p'}{\partial x}\right) - \left(\frac{p'}{L_1^2} + \frac{n'}{L_2^2}\right)\left(\frac{1}{p} + \frac{1}{bn}\right)$$
$$= \frac{1}{D_h p}\frac{\partial p'}{\partial t} + \frac{1}{D_e n}\frac{\partial n'}{\partial t} \tag{4.67}$$

Here, we *can* set $p' \approx n'$, because the term in E that might involve their difference is *not* multiplied by a large number in this case. Indeed, any electric field E in this problem (where there is no source at the end contacts) would arise from the unbalanced charge of $p' - n'$, and so be *at most* of the same order as p' or n'. Hence the entire term in E may be *neglected* here as second order by comparison with the others.

$$\frac{\partial^2 p'}{\partial x^2} - p'\left(\frac{1}{\tau_1} + \frac{1}{\tau_2}\right)\left[\frac{nD_e + pD_h}{D_e D_h(n + p)}\right] = \left[\frac{nD_e + pD_h}{D_e D_h(n + p)}\right]\frac{\partial p'}{\partial t}$$

or

$$\left[\frac{D_e D_h(n + p)}{nD_e + pD_h}\right]\frac{\partial^2 p'}{\partial x^2} - \frac{p'}{\tau_a} = \frac{\partial p'}{\partial t} \tag{4.68}$$

where we have put $1/\tau_a \equiv (1/\tau_1) + (1/\tau_2)$. But the condition of small injection in the present case means that $p' \ll p_o$ *and* $n' \ll n_o$; so

$$\left.\begin{aligned} &(a)\ n + p = n_o + p_o + n' + p' \approx n_o + p_o \\ &(b)\ nD_e + pD_h = n_oD_e + p_oD_h + n'D_e + p'D_h \approx n_oD_e + p_oD_h \end{aligned}\right\} \tag{4.69}$$

which means we can write Eq. 4.68 in a form similar to Eq. 4.55, namely,

$$D_a \frac{\partial^2 p'}{\partial x^2} - \frac{p'}{\tau_a} = \frac{\partial p'}{\partial t} \tag{4.70}$$

in which D_a, given by

$$\frac{1}{D_a} = \left(\frac{n_o}{p_o + n_o}\right)\frac{1}{D_h} + \left(\frac{p_o}{p_o + n_o}\right)\frac{1}{D_e} \tag{4.71}$$

is called the *ambipolar diffusion coefficient.* It has a value approaching D_h for $n_o \gg p_o$ (extrinsic n-type material), $2D_eD_h/(D_e + D_h)$ for $n_o = p_o$ (intrinsic material), and D_e for $p_o \gg n_o$ (extrinsic p-type material).

In other words, the quasi-neutral approximation $|p' - n'| \ll p',n'$, coupled with the extreme condition of small injection $p' \ll p_o$ *and* $n' \ll n_o$, leads to diffusion of the excess-carrier pulse with the ambipolar diffusion coefficient. In the extrinsic case, this diffusion coefficient becomes that of the minority carrier—as we found in our earlier treatments. The fact that the diffusion coefficients for the two carriers are unequal is the essential reason why in less extrinsic cases the pulse of minority carriers may not spread simply with the diffusion coefficient for that carrier. This means that some electric field is developed, in the manner illustrated and discussed in connection with Fig. 4.4*b*. This field has little effect on the minority current, as such, but it does have an appreciable effect on the *difference* between the minority current *inside* the pulse and *outside* the pulse, on account of the drift term $\mu_h p_o E$. The field E inside the pulse will differ from that outside by an amount ΔE proportional to p', and for large enough p_o the resulting *difference* $q\mu_h p_o(\Delta E)$ in conduction current may well compete with the corresponding *difference* in diffusion current (also proportional to p'). It is the *difference* between *current* inside and *current* outside that causes the carrier pulse edge to move, because the motion is a major factor in producing $\partial p'/\partial t$ (see the equation of continuity). Hence, any contribution to $\partial p'/\partial t$ which arises from other than the minority-carrier diffusion currents will cause the pulse edge to move at a rate different from minority-carrier diffusion speeds. Because of the extreme small-injection assumption and the resulting linearity of the problem, it turns out that the entire effect of this induced field can be incorporated artificially into a modified value of "diffusion coefficient for the whole pulse"—the ambipolar diffusion coefficient D_a.

4.3.6 *Concluding Example: The Basis of the Haynes-Shockley Experiment*

Now we are in a position to modify the example illustrated in Fig. 4.7, by the addition of an electric field to the bar. Before the light flash, let there be a direct current I_o throughout the bar, applied by a current source through end contacts located far away from $x = 0$. The situation is like that shown in Fig. 4.2*a*, with the resulting constant electric field, here called E_o, being divergenceless inside the bar and being produced by charges at the end contacts far outside the region of interest around $x = 0$. The current I_o from left to right is accounted for mostly by majority electrons flowing in the *opposite* direction, but producing a current to the right of $q\mu_e n_o E_o$ amps/m^2. There is by comparison very little drift current carried by minority carriers moving to the right, $q\mu_h p_o E_o$, because we shall consider again the *extrinsic* situation in which $p_o \ll n_o$ and μ_e is comparable to μ_h.

Now when the light flash produces excess pairs, the excess minority holes in the resulting pulse are still acted upon by an electric field to the right. If the majority carriers were to "shield" the excess minority carriers completely, in the sense $n' = p'$, the electric field inside the pulse would not be altered from the original value E_o produced by the charges on the end contacts, because there would be no unbalanced space charge to produce a divergence of E. Thus the excess holes would all move to the right with a speed $\mu_h E_o$, in addition to any diffusion spreading of the group that takes place whether the applied field is present or not (the problem is a linear one, to which superposition applies, as long as the injection levels are small enough). On this basis, we might expect a solution for the minority carrier in the present problem to be just that of Eqs. 4.57 and 4.62, except that the entire distribution would also move to the right with a speed $\mu_h E_o$. In other words, x would simply be replaced by $(x - \mu_h E_o t)$.

$$p'(x,t) = \frac{P_o'}{2\sqrt{\pi D_h t}} e^{-t/\tau_h} e^{-(x-\mu_h E_o t)^2/4D_h t} \tag{4.72}$$

The pulse would look just like that of Fig. 4.7, except that it would move to the right with speed $\mu_h E_o$ as it spreads and decays by diffusion and recombination (see Fig. 4.8).

We can back up the solution 4.72 quantitatively if we argue as follows. The electric field in the bar is actually the applied field E_o, produced by the charges at the end contacts, plus a field E' produced by unbalanced charge in the bar, $q(p' - n')$. In the minority-carrier current, we have seen previously that drift terms of the form $q\mu_h pE'$ are negligible compared to diffusion terms $qD_h \partial p'/\partial x$,

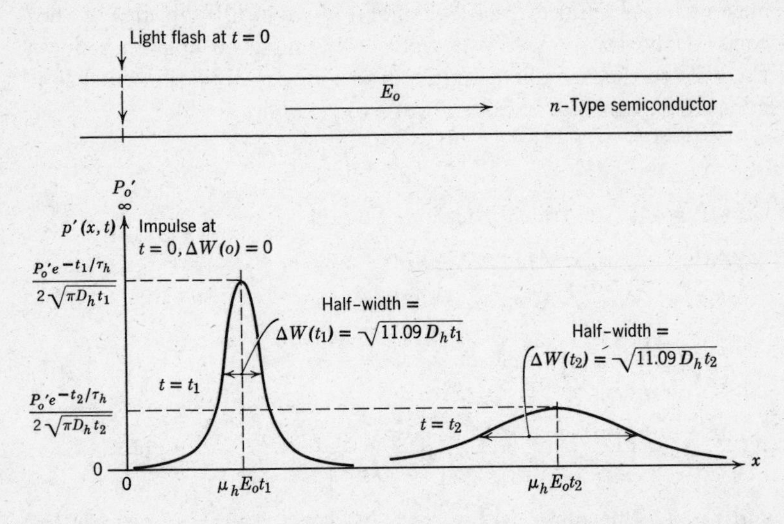

Fig. 4.8. Response of long bar with drift field E_o to impulse light source at $x = 0$, $t = 0$. Assumed: extrinsic bar, small injection conditions, and comparable hole and electron mobilities.

provided the injection level is small (p', $n' \ll n_o$), the bar is extrinsic ($n_o \gg p_o$), and the diffusion coefficients of both carriers are comparable. On the other hand, the applied field E_o may be large—it is certainly not limited in any way by the size of p' or n'. So a drift term of the form $q\mu_h pE_o$ is surely *not* to be treated as negligible compared to the diffusion term. In such cases the minority carrier does *not* move mainly by diffusion, and we may expect from our earlier arguments that the total minority-carrier current is then *not* comparable to the total majority-carrier current. This is obviously the case, because $\mu_h pE_o \ll \mu_e nE_o$ on account of our assumptions that $p \ll n$ and the mobilities are comparable. So, retaining only the necessary drift term, we can write

$$J_h = -qD_h \frac{\partial p'}{\partial x} + q\mu_h pE_o \tag{4.73}$$

and the continuity equation for holes becomes

$$D_h \frac{\partial^2 p'}{\partial x^2} - \mu_h E_o \frac{\partial p'}{\partial x} - \frac{p'}{\tau_h} = \frac{\partial p'}{\partial t} \tag{4.74}$$

inasmuch as E_o and p_o are independent of x and t. If our idea of the basis of the solution 4.72 is right, we should be able to reduce Eq. 4.74 to the form of a simple diffusion-recombination equation by using a change of variable from x to x', where

$$x' = x - \mu_h E_o t \tag{4.75}$$

Carrying out, the transformation, we find

$$\left(\frac{\partial p'}{\partial x}\right)_t = \left(\frac{\partial p'}{\partial x'}\right)_t \left(\frac{\partial x'}{\partial x}\right)_t = \left(\frac{\partial p'}{\partial x'}\right)_t$$

$$\left[\frac{\partial}{\partial x}\left(\frac{\partial p'}{\partial x}\right)_t\right]_t = \left(\frac{\partial^2 p'}{\partial x'^2}\right)_t \tag{4.76}$$

$$\left(\frac{\partial p'}{\partial t}\right)_x = \left(\frac{\partial p'}{\partial t}\right)_{x'} + \left(\frac{\partial p'}{\partial x'}\right)_t \left(\frac{\partial x'}{\partial t}\right)_x = \left(\frac{\partial p'}{\partial t}\right)_{x'} - \mu_h E_o \left(\frac{\partial p'}{\partial x'}\right)_t$$

and Eq. 4.74 becomes

$$D_h \frac{\partial^2 p'}{\partial x'^2} - \frac{p'}{\tau_h} = \frac{\partial p'}{\partial t} \tag{4.77}$$

We already know from Eqs. 4.62 and 4.57 that the solution which meets the necessary conditions at $t = 0$, $x' = 0$ is just

$$p' = \frac{P_o'}{2\sqrt{\pi D_h t}} e^{-x'^2/4D_h t} e^{-t/\tau_h} \tag{4.78}$$

and, in view of the relationship 4.75, we therefore see that our result 4.72 is justified in the present approximations.

The behavior of the excess-carrier pulse discussed in this example illustrates clearly each of the important dynamical features of such carriers. It is particularly the minority-carrier behavior that stands out in the pulse, and if there were a method of detecting these versus time at various positions on the bar, we could observe the drift, diffusion, and recombination of minority carriers experimentally: drift by the speed of motion of the center of the pulse, diffusion by its spreading in width, and recombination by its changes in area. The necessary detector does, in fact, exist in the form of a negatively biased rectifying contact (a form of *pn*-junction diode which we shall study in another place), and the experiment suggested here is a modified form of the ones first performed in 1949–

1951 by Haynes and Shockley at Bell Telephone Laboratories.† These experiments constitute the most direct observations of the effects of minority carriers, and it is worthwhile to become familiar with them to gain conviction in the concept of minority carriers.

4.3.7 *The Effect of Less Extrinsic Material*

If we cannot assume $p_o \ll n_o$ in the problem of Fig. 4.8, we must return to the formulation made in Eq. 4.67, based on the extreme small-injection assumption $n' \ll n_o$ and $p' \ll p_o$. Now, however, whereas the term E' produced by the internal charge unbalance $p' - n'$ is still negligibly small as a coefficient of the second pair of terms, the term E_o may not be so neglected. The argument for setting $p' \approx n'$ was given after Eq. 4.67, and still holds, as does the basis for replacing n and p with n_o and p_o in all coefficients.‡ The resulting equation reads

$$D_a \frac{\partial^2 p'}{\partial x^2} - \mu_a E_o \frac{\partial p'}{\partial x} - \frac{p'}{\tau_a} = \frac{\partial p'}{\partial t} \tag{4.79}$$

with D_a and τ_a given by the development of Eqs. 4.68, 4.70, and 4.71, and

$$\mu_a = \frac{\mu_e \mu_h (n_o - p_o)}{\mu_e n_o + \mu_h p_o} \tag{4.80}$$

being the *ambipolar mobility*. Notice that $\mu_a \rightarrow \mu_h$ for $n_o \gg p_o$ and $\mu_a \rightarrow -\mu_e$ for $p_o \gg n_o$, which are the results we would expect for the drift speeds and directions of minority holes in extrinsic n-type material and minority electrons in extrinsic p-type material, for the *same* direction and magnitude of E_o in both cases. But to find $\mu_a = 0$ for $n_o = p_o$ (intrinsic material) is perhaps a surprise. The effect here is again based on the distinction we have already seen between the motion of a concentration "bump," and the motion of individual carriers in it. The clearest example of this important point is seen in the behavior of the *majority* electrons of

† J. R. Haynes and W. Shockley, *Phys. Rev.*, **75**, 691 (1949) and *Phys. Rev.*, **81**, 835 (1951). See also J. R. Haynes and W. C. Westphal, *Phys. Rev.*, **85**, 680 (1952). An SEEC 16-mm, 30-minute, sound film on these experiments, entitled "Minority Carriers in Semiconductors," has been made by Haynes and Shockley. Pending arrangements for commercial distribution, it may be borrowed for preview or purchased from the Film Librarian, Educational Services, Inc., 47 Galen Street, Watertown 72, Mass. Problem P4.18 employs data from the film.

‡ In really intrinsic material, $n_o - p_o$ and $n' - p'$, which now both occur in the coefficient of E_o, may be comparable; but both are so nearly zero that the main conclusions of interest here will be unchanged.

Fig. 4.8, in the strongly extrinsic case, because the two motions in question are actually *oppositely directed* for these carriers. We know that the pulse of excess holes moves to the right in the field. We also know that $n' \approx p'$ almost everywhere, so there must be a concentration "bump" of electrons *also moving to the right*, in step with the pulse of holes shown. But, of course, the *individual electrons move to the left* in the field!

To understand this, let us forget for a moment about diffusion and recombination, and concentrate only on drift (this is fair, owing to the linearity of the system, and the consequent superposition of the various effects, under small injection conditions). In particular, consider the abrupt uniform concentration bump of Fig. 4.9*a*. Then the only way to get the electron concentration "bump" to move to the right is to have the electron electric *current* (which is always to the *right*) *smaller inside* the pulse than *outside* it. This is equivalent to an electron particle flux (always to the *left*) being smaller inside than outside, so electrons build up on the right-

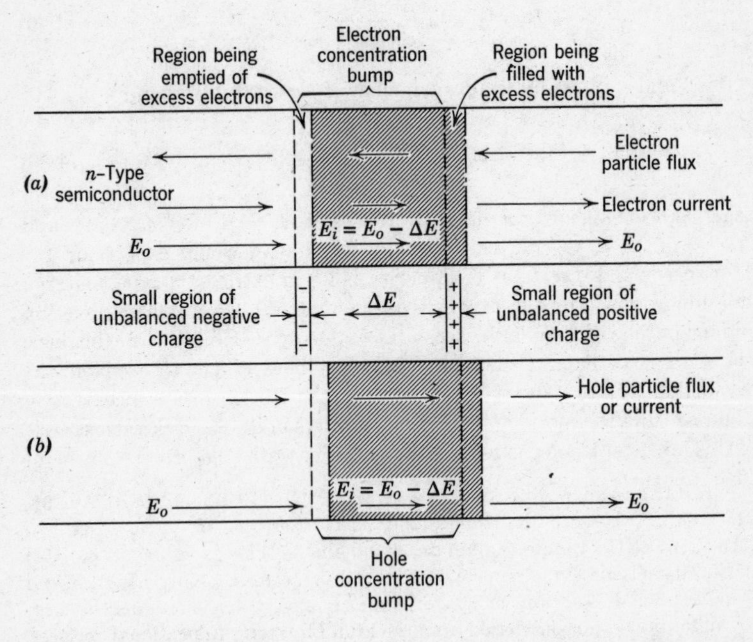

Fig. 4.9. (*a*) Schematic illustrating how inequality of currents inside and outside an excess electron concentration bump can make it move in a direction opposite to that of the individual electrons. (*b*) The corresponding motion of the pulse of holes. Note relative displacement of hole pulse with respect to electron pulse and direction of ΔE.

hand edge and are depleted on the left-hand edge, which corresponds to a rightward motion of the pulse.

Thus the electric field E_i inside the pulse must be *smaller* than E_o outside, because there are more electrons inside, yet less electron drift current there. This difference in field $\Delta E = E_o - E_i$ is proportional to the excess carrier concentration $n' \approx p'$, and results in a difference $p_o \Delta E$ of *minority* hole drift current to the right inside and outside, such that the value inside is *smaller* than that outside (see Fig. 4.9*b*). On the other hand, there is also a minority-carrier drift-current component $p'E_o$, which inside is *larger* than outside ($p' = 0$ outside the pulse in Fig. 4.9). Consequently, the term $p_o \Delta E$ tends to *reduce* the hole current inequality that drives the pulse of holes to the *right*, and for a given injection level p', this reduction is larger as p_o is made larger. This slows down the drift speed of the pulse. Because of the system linearity, this effect can finally be represented as an alteration in the apparent mobility of the holes in the field E_o, leading to Eq. 4.80 for the ambipolar mobility μ_a. In fact, a careful analysis of the continuity relations for Fig. 4.9, neglecting diffusion and recombination, and assigning an unknown drift speed v to both carrier pulses, will lead to the result that $v = \mu_a E_o$ from the two simultaneous continuity equations. We may assume $p' \approx n'$ and neglect second-order products like $p' \Delta E$ and $n' \Delta E$ (but not $p_o \Delta E$ or $n_o \Delta E$). This is a worthwhile example to go through in detail one's self to understand this whole ambipolar issue. It is taken up in problem P4.20.

Finally, we can see from Fig. 4.9 also that the reduction ΔE of the electric field inside the pulse must come from charges at the edges of the pulse, and that these must result from a small relative displacement between the hole and electron pulses in the direction shown. We may imagine this displacement to arise from an initial tendency of electrons to move left and holes to move right.

PROBLEMS

P4.1 Show that the equations of continuity for the separate carriers (Eqs. 4.9 and 4.10) are consistent with the general equation of conservation of charge

$$\nabla \cdot \mathbf{J} = -\frac{\partial \rho}{\partial t}$$

P4.2 (*a*) Write the continuity equations analogous to Eqs. 4.13 for the case of extrinsic *p*-type material.

(*b*) Consider material for which p_o and n_o may be comparable to each other, but $n' \ll n_o$ *and* $p' \ll p_o$. Write continuity equations like Eqs. 4.13 for this material. See Sec. 1.6.2.

P4.3 Carry out the analysis of the field for the three sheets of charge in Fig. 4.2*b*, as suggested on p. 148, and verify the conclusions stated there.

P4.4 A bar of conducting material of length d, with uniform dielectric permittivity ϵ has perfectly conducting end contacts. The conductivity of the material is

$$\sigma = \sigma_o \left(\frac{2d}{x + d} \right)$$

where x is the coordinate along the bar.

(*a*) Find the electric field in the bar when a voltage V_o is applied between the contacts. Neglect fringing.

(*b*) Find the charge density in the bar, and the surface charges on the contacts. Show that the whole system is electrically neutral.

P4.5 (*a*) Consider a condition of thermal equilibrium in a one-dimensional graded bar with known net doping $N(x)$ (see Fig. 3.3, for example). Introduce the new variable u of Eq. 4.21*a*, and find the differential equation for u in the form

$$L^2_{Di} \frac{d^2u}{dx^2} = \sinh u - \overline{N}$$

Identify $\overline{N}$ and evaluate the "intrinsic Debye length" L_{Di} in terms of the given parameters. Show that L_{Di} may be written in the form

$$L_{Di} = \sqrt{\left(\frac{D_e + D_h}{2} \right) \tau_{di}}$$

where $\tau_{di} = \epsilon/\sigma_i$. Compute its numerical value for Ge and Si at 300°K.

(*b*) Suppose we have a graded semiconductor with

$$\overline{N} = \overline{N}_o ax; \quad \overline{N}_o > 0; \quad a > 0$$

Find and sketch an approximate solution for ψ that arises from charges within the bar, has a field that vanishes at $x = \pm \infty$, but is nevertheless based on the approximation that the bar is everywhere electrically neutral. Find and sketch the corresponding approximate electric field, and from it an approximation to the charge density ρ which was neglected.

(*c*) Evaluate the ratio ρ/qN from your approximate answer to part (*b*), and show that its maximum value is negligible if the change in $\overline{N}$ over one intrinsic Debye length L_{Di} is small.

P4.6 Consider a homogeneous semiconductor bar in which, under all conditions, rnp is the number of hole-electron pair recombinations taking place per m³ per sec, with r a constant. Let p_o and n_o be the equilibrium carrier concentrations. Evaluate G_{th} for the bar.

P4.7 Suppose the bar of P4.6 is *uniformly* illuminated by penetrating light that produces a constant number g_L hole-electron pairs per m^3 per sec. Find the steady-state values of p and n. Assume the problem is one-dimensional.

P4.8 Suppose next that the bar of P4.6 is at thermal equilibrium, and the light source of P4.7 is suddenly turned on at $t = 0$.
(*a*) Work out the build up of excess carrier concentration as a function of time. Are there any acceptable solutions for which $p' \neq n'$?
(*b*) Show that for small injection, a lifetime can be defined which describes correctly the recombination process for excess (minority) carriers.

P4.9 The bar of P4.7 is in the steady state when, at $t = 0$, the light is extinguished.
(*a*) Find the decay of excess carriers as a function of time and compare with the build-up found in P4.8.
(*b*) Repeat for small injection and check the concept of a lifetime.
(*c*) Plot carefully on semilog paper the result of (*a*) for large injection, and compare it with a similar plot for small injection conditions with the *same* initial excess at $t = 0$ for both cases.

P4.10 Solve P4.8 and P4.9 with conditions of the kind described in P4.2*b*.

P4.11 Solve the problem of Sec. 4.3.2 for a p-type bar, with $D_e = 2D_h$.

P4.12 Consider the region $x > \delta/2$ in P4.11, and let $x' = x - \delta/2$.
(*a*) Show that $n'(x')/n'(0)$ is the probability that an excess electron at $x' = 0$ will reach *at least* x' before recombining.
(*b*) Next, find the (differential) probability that an excess electron at $x' = 0$ will reach x' and then recombine in the next small interval dx'.
(*c*) From this, find the average distance $\langle x' \rangle$ an excess carrier goes in its lifetime.
(*d*) Find $\langle x'^2 \rangle$.

P4.13 The homogeneous n-type semiconductor bar AA' in Fig. 4.10 is center-tapped for use as a photoconductor in the bridge circuit shown, similar to the one employed in the laboratory experiment of Sec. A3.3.1. Without light, the open-circuit voltage ΔV is zero. It is desired to determine the proper choices of thickness δ and source voltage V_o to maximize the unbalance signal power available from the terminals at ΔV, for a given steady light intensity.

Assume that the light penetrates into the semiconductor a distance which is negligibly small compared to both the diffusion length and the bar thickness δ, so that it produces m excess carrier pairs per unit area per unit time on the illuminated surface of the bar. Consider the problem as being one-dimensional, with the only significant variations taking place in the thickness dimension δ.

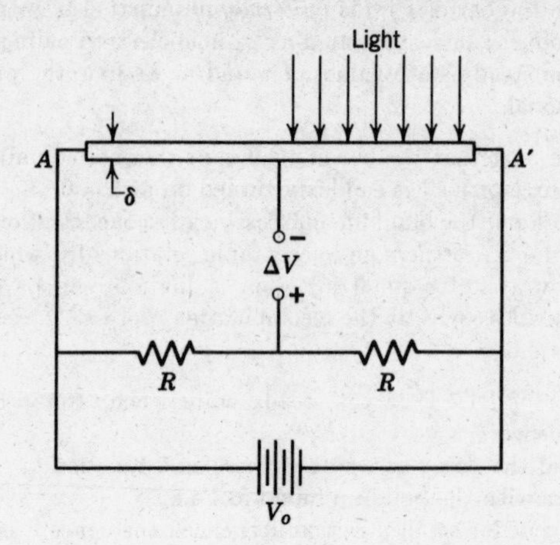

Fig. 4.10.

Treat only the case of small injection, and include a lifetime τ in the bulk *and* an *infinite* "surface recombination rate" (i.e., $p = p_o$, or $p' = 0$) on the *unlit* surface.

(*a*) Find the distribution of excess carriers as a function of position across the bar.

(*b*) Find the change of conductance ΔG of the illuminated half of the bar, if the whole bar is $2l$ long and W wide. Sketch ΔG vs δ, with everything else fixed.

(*c*) Find the signal power available from ΔV as a function of ΔG, the *unlit* bar conductance G_o, and the voltage V_o. Then re-express the result in terms of the dc power dissipation P_o in the unlit bar, instead of V_o.

(*d*) What first prevents use of arbitrarily large dc values of V_o on a given bar?

(*e*) What conditions will limit the choice of V_o and δ to obtain the largest possible available signal power from ΔV per unit light intensity, with all other dimensions fixed?

(*f*) Assume for intrinsic Ge a simple exponential dependence of G_o on the temperature rise above room temperature, with an exponent of 6% per °C, and a temperature rise of the bar of about 10°C per watt of dc power dissipated in it. Find the optimum power dissipation for a given bar, and in particular evaluate the corresponding battery voltage V_o in the case of a bar having 1000 ohms resistance at room temperature.

(*g*) Repeat parts (*c*) through (*f*) if ΔV and the battery are interchanged in position. Comment on the comparison with the arrangement of Fig. 4.10. Which arrangement uses more power from the *battery*, assuming that the unbalance signal power has been maximized in each case by choice of δ and V_o?

P4.14 Derive Eq. 4.60 by direct integration of Eq. 4.55.

P4.15 Show for the case of pure diffusion (Eq. 4.59) that

$$<x^2> \equiv \frac{\int_{-\infty}^{\infty} x^2 p'_\infty(x,t)\,dx}{\int_{-\infty}^{\infty} p'_\infty(x,t)\,dx} = 2D_h t$$

Do this by direct integrations (without solution) of Eq. 4.59, suitably multiplied. Use integration by parts. Check the result by carrying out the integrations, using the solution Eq. 4.62. Compare all this with the reasoning in P4.12(*c*) and (*d*). Then give a new interpretation to the Debye length, $L_D = \sqrt{D\tau_d}$ (see Eqs. 4.25 and 4.26, and also P4.5 and P4.12).

P4.16† Consider *adding* Eqs. 4.65 and 4.66 to find out something about the way $p' - n'$ behaves. Solve the resulting equation in the steady state for a case like $x \geqslant \delta/2$ in Fig. 4.4, and interpret the result.

P4.17 Show that, neglecting diffusion, the solution to Eq. 4.74 (or Eq. 4.79) is

$$p'(x,t) = e^{-t/\tau} f(x - vt)$$

where $v = \mu E_o$ and $f(x)$ is the initial distribution pattern of $p'(x,0)$.

P4.18 The following data are taken from a modified form of the Haynes-Shockley experiment‡ in which a detector of holes is placed on an n-type Ge bar of 5 ohm-cm resistivity similar to the one illustrated in Fig. 4.8. The pulse received by the detector is observed as a time function on an oscilloscope. The distance W between light source and detector is varied mechanically by moving the bar and detector with respect to the light source. Recorded for each distance setting are the delay time t_d between light source flash and center of received pulse, the peak deflection V_p of the received pulse, and the time duration t_p between places where the pulse is $V_p/2$ in height.

The measured voltage drop between open-circuited probes on the bar 1.25 cm apart was 9.70 volts. The bar temperature was 31°C with the voltage applied.

† This problem requires more than the minimum background.

‡ These data were obtained from the experiment performed in the film referred to on p. 181.

W(mm)	t_d(μ sec)	V_p(detector, millivolts)	t_p(μ sec)
1.0	9.0	380	6.9
1.5	13.0	310	
2.0	16.5	280	8.9
2.5	20.0	200	
3.0	24.0	160	10.0
3.5	28.0	120	
4.0	31.5	120	11.2
4.5	36.0	100	
5.0	39.5	95	11.8
5.5	43.5	80	
6.0	47.0	77	13.0
6.5	51.0	70	
7.0	55.0		
7.5	59.0		
8.0	63.0	Signal shape hard to see in this range.	
8.5	67.0		
9.0	70.5		
9.5	75.0		

(*a*) Find the mobility and diffusion coefficient for holes in this bar. Check the Einstein relation.

(*b*) Show from the theory that the number of holes in the pulse at any time t_d is proportional to the peak height $p_{\max}$ times $\sqrt{t_d}$. Argue that this is very nearly the same as being proportional to $V_p\sqrt{t_d}$, and by using this fact determine the lifetime of holes in the bar. Show that the number of holes in the pulse at time t_d should also be proportional to the product $p_{\max}W$, which in turn is very nearly the same as being proportional to $V_p t_p$. Use this fact to determine the lifetime of holes. Discuss the agreement between the two methods.

(*c*) If we were able to measure a minimum detector signal of 5 millivolts in the V_p column of data, what maximum delay could we observe? How long would the bar have to be to observe this delay with the electric field used here?

(*d*)† Estimate corrections, especially in the Einstein relation, which might be made to account for ambipolar effects. Note that the equilibrium minority-carrier concentration increases very rapidly with temperature.

(*e*) Derive from your mobility value the value at room temperature, using the known temperature dependence of μ_h.

(*f*) The data for V_p were reproducible—the irregularities in the received signal at different values of W occurred repeatedly just as given. Suggest reasons for such behavior in this form of experi-

† This problem requires more than the minimum background.

ment, where the light spot is moved with respect to the bar to vary W.

(*g*) What effect does jitter in the light source intensity from one pulse to the next have on the determination of D, assuming we take single-trace 'scope photographs to measure half-width? Answer the same question for the determination of τ. Estimate the expected scatter of the data for determining D, if the half-widths are 3 per cent accurate from 'scope photographs.

(*h*) The source light pulse is neither infinitely thin nor of zero time duration. Consider separately what effect these two defects will have on the experimental results. Estimate the quantitative effect if actually the light pulse was about 1 μs in duration, and 1/10 mm wide.

(*i*) Consider the effect of having the light pulse infinitely thin and of zero time-duration, but applied to the top surface of the bar only (e.g., it does not penetrate). What effect will this "side illumination" have on the experiment? (Note that a two-dimensional form of P4.21 may be helpful; but remember that there is actually an electric field applied along the bar).

(*j*) In reality, *all* three light-source defects mentioned in parts (*h*) and (*i*) above were present *together* in the experiment. How then can we apply the "one dimensional," "line-impulse source," analysis at all, and how should the observed data be handled to account for these effects? What was done about these difficulties in the film referred to above?

P4.19 Consider Eq. 4.74 (or 4.79) for excess hole concentration with e^{st} time variation. That is, let $p'(x,t) = P'(x,s)e^{st}$.

(*a*) Find the differential equation for $P'(x,s)$. Suppose we have an arrangement similar to Fig. 4.8, except that the light flash is replaced by a small steady "light bias," on top of which is a smaller sinusoidal variation of light intensity. Suppose we also have a means for detecting minority holes directly at some *fixed* distance W_o down the bar.

(*b*) Consider first the solution $P'(x,0)$ for $s = 0$. Sketch $P'(x,0)$ versus x for $E = 0$, and then for $E = E_o$.

(*c*) Determine the transfer function between the light modulation and the detector response, for small injection conditions and $s = j\omega$. Then consider using this arrangement to provide the feedback for a sinusoidal oscillator, in which case a phase shift of π is necessary in the transfer function.

(*d*) Find the frequency at which the phase shift of the transfer function of part (*c*) is π, and the corresponding attenuation of the excess hole concentration. Do this in two limiting cases: first, neglect entirely the effect of diffusion; then neglect entirely the drift field E_o. Finally, determine the conditions under which it is fair to neglect the effect of diffusion compared to drift.

(*e*) Make a sketch of the attenuation and phase versus frequency for the transfer function.

P4.20† As suggested in the text on p. 183, carry out the direct analysis of the continuity relations for Fig. 4.9, neglecting both diffusion and recombination. Pay careful attention to the values of the electric field inside and outside the pulse, and show that $v = \mu_a E_o$.

P4.21† Consider the continuity equation for minority holes in three dimensions (Eq. 4.13a), for the case $g = 0$.

(*a*) Assuming that the holes flow only by diffusion (vector form of Eq. 4.29), find the resulting differential equation.

(*b*) Let $\tau_h = \infty$ in the result of part (*a*), and consider the problem of the spread of a "point source" of holes—that is, an impulse at $x = y = z = 0$, $t = 0$ containing P_o' excess holes total. Show that if r is the radial coordinate,

$$p'(r,t) = \frac{P_o'}{(4\pi D_h t)^{3/2}} e^{-r^2/4D_h t}$$

is the required solution (use rectangular coordinates to substitute in the differential equation, and to do the required integrals, which are tabulated in Eqs. 3.17 and 3.18).

(*c*) Evaluate $<r^2>$ as a function of time (see also P4.15).

(*d*) How must the above solutions be modified if τ_h is a finite constant?

† This problem requires somewhat more than the minimum background.

APPENDIX
Laboratory Experiments

GENERAL DISCUSSION

A1.0 INTRODUCTION

The general model of semiconductors developed in the text rests on a firm experimental foundation; in fact, as a result of the extensive investigations stimulated by the development of the transistor, our understanding of the physics of germanium is more nearly complete than for any other material. The first four experiments selected here are intended to supply the experimental foundation for much of the text material in this book. The fifth experiment deals primarily with matters treated in another volume (PEM).

There are four main features of the semiconductor model which we will investigate experimentally in the first four experiments:

1. There are two types of charge carriers (conduction electrons and holes) which contribute to the conduction processes in a semiconductor.
2. The concentrations and mobilities of the charge carriers depend on temperature in a manner explicable by the theory.
3. Electron-hole pairs may be produced thermally, electrically, or optically by supplying an "activation" energy (the gap energy), and the conduction electrons and holes so produced contribute independently to the conduction process.

4. Excess carriers in a semiconductor display the dynamical characteristics of drift, diffusion, and recombination.

More than one of these features may be involved in any one of the experiments.

These experiments are based on much more elaborate ones which underlie our understanding of semiconductors. For our purposes, the equipment has been simplified to permit student construction and preparation and, at the same time, to maintain an accuracy sufficient to obtain meaningful results which may be compared with the theory presented in the text. Those who wish to pursue this type of measurement further, can find descriptions of more elaborate and precise equipment throughout the literature.

For the sake of brevity, access to enough information on electrical circuits and measurements has been assumed in writing these laboratory notes to make unnecessary much discussion or analysis of circuits per se. Sample preparation and properties peculiar to semiconductors have, however, been treated much more explicitly.

Whereas circuits for some of the required measuring equipment have been included, e.g., a differential amplifier and a pulse generator, the appreciable time required for building and testing these circuits may, of course, be eliminated if equivalent commercial equipment is available. However, in the general scheme of our whole objectives, the transistor circuit work involved in "doing it oneself" has positive value.

A1.1 TIME REQUIRED

The time required for the experiments varies a good deal, depending on the dexterity of the student and whether samples break, etc. Fairly generous estimates of the time required for various parts of the experiments are:

Conductivity

Sample preparation and mounting—2 hours.
Low-temperature conductivity—2 hours.
Oven construction and high-temperature conductivity—3 hours.

Hall effect

Construction of differential amplifier—1 hour.
Mounting of Hall-effect samples—1 hour.
Hall-effect measurements—2 hours.

Energy gap

Spectrometer calibration and measurement of energy gaps—3 hours

Drift mobility

Construction of pulse generator—1 hour.
Sample polishing—1 hour.
Sample etching and mounting—1 hour.
Measurement of drift mobility—3 hours.

Manufacture of a simple diode

Tinning, polishing, and etching of first wafer—½ hour.
Alloying, cleanup, and mounting of first diode—1 hour.
Electrical evaluation of first good diode—1½ hours.
Allowance for rejects—1½ hours.
Additional study and comparisons with point contacts—3 hours.

A1.2 EQUIPMENT LIST

It is assumed that common electrical components are available; e.g., resistors, capacitors, common transistors, batteries, and meters.† Equipment which is peculiar to these experiments is listed below. Items marked with an asterisk (*) are used in more than one of the experiments, and are listed only under the experiment in which they are first used.

1. Conductivity
   - Germanium samples* (see p. 195)
   - Polyflux or other suitable flux* (see p. 196)
   - #26 stranded wire with Teflon insulation* (see p. 196)
   - 60% lead-40% tin coreless solder
   - Fine emery paper*
   - 150-ml beaker*
   - Dry ice
   - Acetone*
   - Electrical tape* (see p. 203)
   - Aluminum foil
   - Power resistor (see p. 201)
   - Variable power transformer* (optional in this experiment)

† A multimeter, a plastic chasis, a 12-volt transformer, and a sufficient selection of components, can be assembled together for under $50.00, including a carrying case. An oscilloscope is *needed* in only one experiment here.

Soldering iron (small tip, pencil-type)*
Wooden spring-type clothespins*
Thermometers* (see p. 200 and p. 203)

2. Hall effect

Permanent magnets (Magnetron magnets will do; fields of 2000 gauss and 4000 gauss are desirable.)
Wooden dowel (¼ in.) or wooden pencil
Rubber bands*
Coarse sandpaper
Variable voltage source* (optional)
Matched silicon transistors (see Fig. A6)
Carbon potentiometers* (see Fig. A8)

3. Energy gap

Samples and spectrometer (see p. 215)

4. Drift mobility

Linde-*A* lapping powder and metallurgical lapping paper*
Glass plate*
3% USP hydrogen peroxide*
Hot plate (up to 200°C)*
Cork

5. Manufacture of a simple diode

n-type 1-5 ohm cm Ge wafers, 100 × 100 × 10 mils, (see p. 195)
Indium spheres 30 mils in diameter (see p. 234)
Coreless solder (see p. 232)
Variable power transformer
Alcohol or toluene or acetone*
Nokorode Soldering Paste (see p. 234), or NH_4Cl in ethylene glycol
Plastic clothespin

MAJORITY-CARRIER EFFECTS

A2.0 INTRODUCTION

The dc conduction properties of a homogeneous sample of a semiconductor depend on the concentrations and mobilities of the charge carriers. In extrinsic material, investigation of the conduction properties gives information about the *majority* carriers; in intrinsic material, we obtain information about the combined effects of conduction electrons and holes. Two relatively simple experiments which yield important information are: the measurement of conductivity as a function of temperature; and the measurement of the Hall effect.

A2.1 SAMPLE PREPARATION

The preparation of the samples is the same for the conductivity and Hall-effect experiments. The samples consist of two bars of germanium, one n-type and one p-type.† The room temperature resistivity of the bars is in the range of 1 to 3 ohm-cm. The dimensions are about 1 cm x 1 mm x 1 mm (the actual dimensions will be needed in the experiments and should be measured carefully).

In both experiments, we will be passing current through the bar and will need electrical contacts on each end of each bar. In the conductivity experiment, we will deduce the magnitude of the electric field in the bar from measurements of the voltage drop between the two end contacts. Therefore, we need low resistance contacts to render negligible the voltage drop across them. We will proceed by tinning the ends of the sample, cleaning the long surfaces, and attaching lead wires to the ends.

Soldering to semiconductors is not always a simple matter (germanium is much easier to solder than silicon). Roughen the ends of the bars with fine emery paper; be gentle—germanium is brittle. Arrange a method to hold the sample horizontally so that solder and flux will not run onto the long surfaces. The end to be soldered should be easily accessible. (A wooden spring-type clothespin works very well.) The solder to be used must not weaken at the highest temperature attained in the experiments (180°C). 60% lead-40% tin is acceptable; 50%-50% softens between 180°C and 190°C; 60% tin-40% lead, which is the usual "electronic" type of solder, is *unsatisfactory*. It is preferable to use a solder without a flux core. (If a flux core solder is used, the flux should be

† A set called the AAPT set, consisting of a germanium p-type bar, an n-type bar, and a $\frac{1}{8}$ in. square n-type wafer, may be purchased for about $1.00 from Semimetals, Inc., 172 Spruce St., Westbury, L.I., N. Y.

allowed to burn away on the soldering iron before applying the molten solder to the germanium.)

Using a piece of solder or other applicator, apply a drop of Polyflux† to the end surface of the germanium. Put some solder on the tip of a clean, well-tinned soldering iron and apply the molten solder to the germanium end surface. Leave the iron in contact with the solder for several seconds until the germanium is well heated. Try to avoid getting solder on the long faces of the germanium. After the solder cools, inspect the end of the bar; if the solder does not cover the entire end surface, repeat the procedure.

If solder has run over onto the long surfaces of the bar, remove it by polishing lightly on fine emery paper laid on a flat surface. Use gentle pressure in this operation and polish only until the excess solder is removed.

With the end surfaces of both bars tinned, attach 8-in. lengths of No. 26 stranded hook-up wire to both ends of each bar. It is preferable to use Teflon insulated wire; ordinary plastic insulation will soften and flow at the highest temperatures attained in this experiment. (The experiment can be done with plastic-covered wire, but it is then critical that the highest temperatures be maintained for only a short time.) The wire is attached easily by placing a drop of Polyflux on the tinned germanium surface, touching the wire to the surface and heating briefly with an iron. The contacts made in this way should be fairly strong—in testing them, pull gently to avoid breaking the germanium. If a contact breaks off, the tinning procedure was probably at fault—repeat the steps above, giving careful attention to cleanliness. If the sample has been handled extensively before soldering, it may be necessary to clean it in alcohol or acetone.

Metal-to-semiconductor contacts, such as we have just described, are not always "ohmic" and of low resistance; they may be rectifying. As a preliminary check on your contacts, measure the resistance of your sample with an ohmmeter; check to make sure the resistance is independent of the direction of current flow. Calculate the resistance of your bar from the dimensions and the expected resistivity (between 1 and 3 ohm cm at room tempera-

† Polyflux is made by Industrial Craftsmen, Inc., 261 Old Billerica Road, Bedford, Mass. Using a proper flux is essential.

ture). If there is any indication of poor contacts, they should be removed and resoldered.

A2.2 DETERMINING CONDUCTIVITY TYPE

It is often convenient to have a means of determining whether a sample is p-type or n-type; the Hall effect to be described later is one means of doing this, but it requires a good deal of sample preparation and equipment. A simple, convenient test makes use of the thermoelectric properties of semiconductors.

Set your multimeter on its most sensitive voltage range (0.25 volt). Heat one multimeter probe with a soldering iron or a match flame for a few seconds. Touch both probes to the germanium sample and observe the meter deflection. If the hot-probe voltage is positive with respect to the cold probe, the sample is n-type; if the voltage is negative, the sample is p-type. Since the labels on the samples are not always correct, you should make this test before proceeding.

Essentially, what you have done is to make a thermocouple with the germanium sample as one leg of it. The detailed theory of thermoelectric effects is rather complicated, but we can indicate their origin by using our simple model of a semiconductor. The distribution of thermal velocities of the charge carriers in a small region of the semiconductor will depend on the temperature of that region. If a temperature gradient exists in the sample, the distribution of velocities will vary with position, carriers near the hot probe tending to have higher velocities than those near the cold probe. In thermal equilibrium, no net current flows because thermal velocities balance; in a temperature gradient, this balance is destroyed and a current can flow. This current is proportional to the temperature gradient.

In the measurement described above, the actual current drawn by the meter is much less than that which could be supplied by the sample (i.e., it is a nearly open-circuit measurement). In this case, there must be an internal electric field in the sample which produces a drift current to just compensate the current arising from the differences in thermal velocities. This field, produced by small deviations from charge neutrality, is proportional to the temperature gradient. Thus the measured voltage is proportional to the temperature difference between the probes.

The sign of the voltage depends on the sign of the charge carriers, because the electric field must in every case be in a direction to drive carriers *toward* the hot probe. In a more complete theory, the details of the scattering of carriers also plays a role.

This would seem to be the first experimental verification you have made that two types of carriers exist in semiconductors. However, the fact that both positive and negative thermoelectric voltages are observed in metals which have only one type of carrier makes the conclusion somewhat dubious. (The model used to describe conduction in metals is rather different from that for semiconductors because of the higher concentrations of conduction electrons.) If you accept the fact that your p-type and n-type samples differ only in the impurity type, and that the concentration of these impurities is very small on a normal metallurgical scale, then the evidence is a good deal stronger. Differences in the sign of thermoelectric voltages in metals are observed for different metals, but not in the same metal when the kind and amount of impurity is altered slightly.

A2.3 MEASUREMENT OF CONDUCTIVITY

A2.3.1 *General Arrangement*

The conductivity is defined from the equation

$$\mathbf{J} = \sigma \mathbf{E} \qquad \text{(See Eq. 1.30)}$$

and in terms of the charge carrier concentrations and mobilities

$$\sigma = nq\mu_e + pq\mu_h \qquad \text{(See Eq. 1.31)}$$

Thus by measuring the conductivity as a function of temperature we may hope to get information about the carrier concentrations and mobilities as functions of temperature.

In measuring the conductivity, we do not measure current densities and electric fields directly; we measure total current and voltage. In a thin bar of length l, with ideal end contacts, the current density is constrained by the geometry of the bar to flow along the bar; and if the material is homogeneous and the contacts are uniform, the current density is constant across the cross section A. Thus the total current is

$$I = JA$$

The voltage drop across the sample is obtained by integrating the electric field, which is directed along the long sample axis and does not vary with position.

$$V = El$$

Thus the conductivity is

$$\sigma = \frac{J}{E} = \frac{I}{A} \cdot \frac{l}{V}$$

It should be clear that in this type of measurement we rely directly on the assumption of ideal end contacts and a homogeneous bar. A suitable circuit for the measurement is shown in Fig. A1*a*.

We can make the measurement less sensitive to the character of the end contacts by making a four-terminal measurement as shown in Fig. A1*b*. With a high-impedance detector, no current flows through the voltage measuring probes, so the resistance of these contacts does not affect the measurement. If the voltage measuring probes are far from the end contacts, the current density will be uniform between them regardless of the end-contact characteristics. We do not use this method because a satisfactory temporary attachment of the voltage-measuring leads is not so easy to arrange, and we wish to avoid a more permanent pair of leads because we shall want to use the same sample for later measurements without them.

A2.3.2 *Sample Mounting and Temperature Control*

The sample temperature is to be measured by using a thermometer; thus the sample must make good thermal contact with the thermometer bulb. Sample and thermometer must be placed in an environment in which the temperature may be controlled. Simple arrangements which achieve this objective are described below. Other methods of achieving the temperature range desired are, of course, possible with more specialized equipment. To see clearly both intrinsic and extrinsic effects, the temperature range should extend from about 180°C down to 0°C or below.

It is probably easiest to start with the low temperature measurement. The low temperatures are obtained by using a cooling bath; the lower the minimum bath temperature, the more meaningful the data. Dry ice and acetone† gives a temperature of about −70°C; if this is not available, ice and water (0°C) may be used. For temperatures below −20°C, a mercury thermometer is not suitable; use an alcohol or a toluene thermometer.‡

† Acetone is flammable; observe reasonable safety precautions.

‡ Toluene thermometers are less expensive but do not cover the same range as alcohol thermometers.

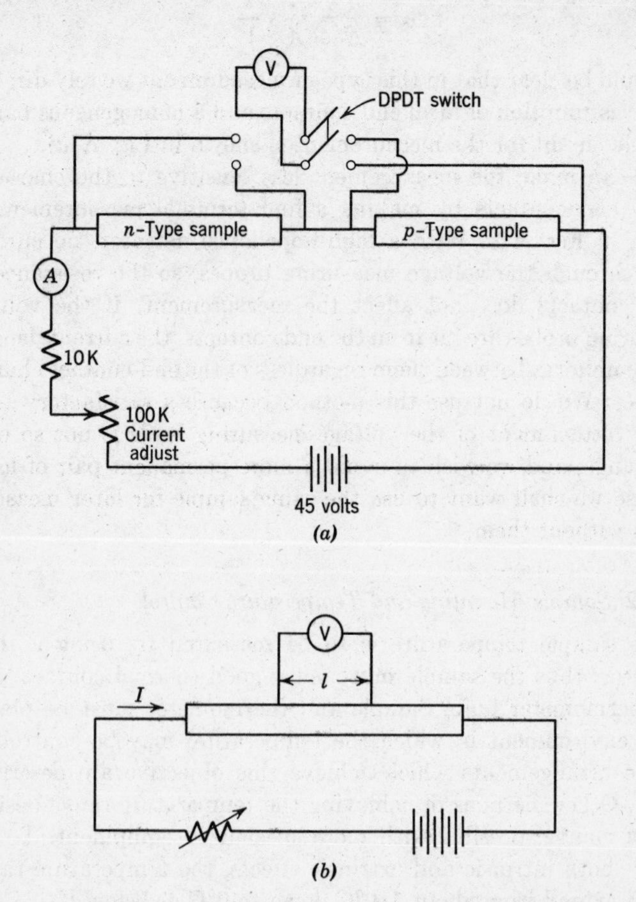

Fig. A1. Arrangements for conductivity measurements.

Attach your samples by taping one lead of each sample to the thermometer; the samples should be in contact with the thermometer bulb. Insulation of the samples and leads from the cooling mixture is not required. Suspend the samples and thermometer near the center of a 150-ml beaker of the cooling mixture. A convenient method of suspension is shown in Fig. A2. Before cooling the samples, measure the conductivity at room temperature.

Measurements should be taken at about 5° intervals as the samples and cooling bath warm from the minimum temperature. If dry ice is used, take measurements up to about 0°C. Keep the amount of dry ice in the cooling bath to a minimum so that warming will occur in a reasonable time. If ordinary ice is used, measurements must be taken up to about room temperature.

For the high-temperature measurements, we use a 50-watt tubular ceramic wire-wound power resistor as an oven. The inside diameter of the resistor is $\frac{1}{2}$ in. and it is about $6\frac{1}{2}$ in. long. A resistance value of 50 ohms is suitable. Power is delivered to the

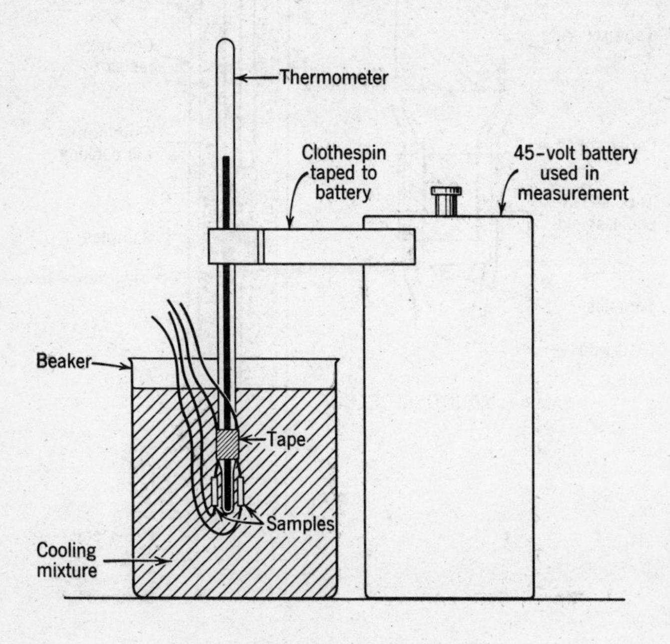

Fig. A2. Arrangement for cold bath.

resistor by using a variable transformer supplied from the 115-volt line. About 40 to 50 volts are required to maintain 180°C within the oven, depending on the oven construction. (If a variable transformer is not available, a suitable series resistance may be used.)

The samples are mounted as shown in Fig. A3. It may be convenient to make measurements on both the *p*-type and *n*-type samples during the same temperature run. Hold the samples against the thermometer bulb (check that the thermometer covers the

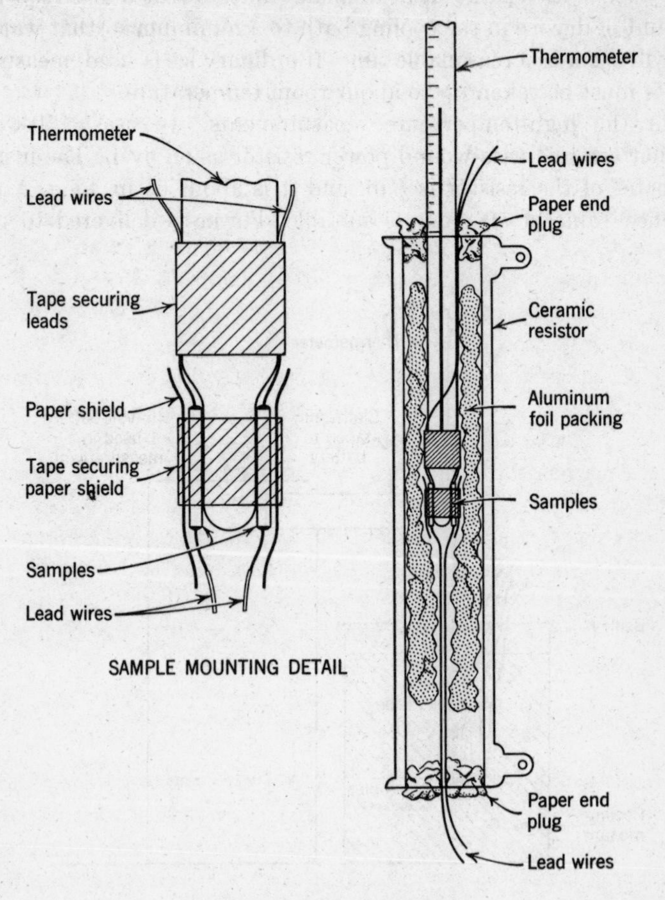

Fig. A3. Arrangement for oven.

desired temperature range) and tape one lead of each sample to the thermometer close to the sample (black vinyl electrical tape does work, although it softens at the highest temperatures). Wrap the samples and thermometer with a piece of paper for insulation; secure the paper with tape. Using aluminum foil whose width is somewhat shorter than the oven, wrap the assembly with several layers of foil until it is a snug fit in the oven. The assembly will be easier if the aluminum foil is wrapped with its diagonal along the thermometer, rather than with an edge along the thermometer. (Don't make it so tight that you break the thermometer or samples when inserting them.) The samples and thermometer bulb should be near the center of the aluminum foil and the oven. After inserting the samples in the oven, with the leads extending from each end of the oven, plug the ends of the oven with paper to cut down air currents through it. Check for electrical continuity and undesired short circuits before proceeding.

Measurements should be taken by first heating the oven and samples to 180°C and then taking measurements as the system cools. Readings should be taken about every 5° down to about 55°C (below this temperature cooling is very slow). When heating the oven, remember that the temperature indicated by the thermometer lags behind that of the resistor, and that the temperature of the samples will continue to rise after oven power is turned off. Heat the oven to about 165°C, turn off the power, wait until the temperature has stabilized and then heat slowly to 180°C. Reduce the power supplied to the oven. The cooling rate should be about 1.5°C per minute; it may be controlled by varying the amount of power supplied to the oven during cooling. Try to complete the measurements at the highest temperatures as quickly as possible since this method of sample mounting is not ideal for high-temperature use.

A2.3.3 *Other Experimental Precautions*

In this and later experiments, we are often assuming a functional form (such as Ohm's law) and evaluating a parameter which occurs in the function (such as the conductivity). An obvious requirement is that the functional form be correct if we are to ascribe any meaning to the parameter we measure. An experimental check of the validity of the functional dependence of measured quantities can serve the dual purpose of pinpointing experimental difficulties and of defining usable ranges of the variables. In this experiment, the experimental verification that the voltage across the sample is linearly related to the current flowing through it serves to verify that the contacts are suitable and that heating of the sample by the current is not excessive. In dealing with new or unfamiliar materials, this type of check is essential.

A2.3.4 *Summary of Experimental Procedure*

1. Check conductivity type with hot probe.
2. Solder leads to samples.
3. Mount samples and measure conductivity at low temperature.
4. Remount samples and construct oven; measure conductivity at high temperature.
5. Measure sample dimensions.

A2.3.5 *Analysis of Data*

1. Using the model of semiconductor behavior developed in the text, show that at sufficiently low temperatures the temperature variation of the conductivity arises from the variation of the majority carrier mobility with temperature. Referring to Table 1.2, decide how to plot your data in order to make a comparison with the results given there. (In deciding how to plot data, a choice of new variables which, when used as ordinate and abcissa, should lead to a straight-line relationship allows an easy check on both the functional form assumed and the values of parameters.) Estimate the accuracy of your results and make the comparison with Table 1.2.

2. Show that, at high temperatures, the temperature variation of the conductivity arises mainly from changes in carrier concentration. Decide how to plot your data to obtain a value of the energy gap. Estimate the accuracy of your value, considering possible meter errors, and compare with Table 1.1.

3. Using the values of mobilities given in Table 1.2 and the value of n_i given by Eq. 3.59*a*, find the impurity concentrations in your samples.

4. Find the temperatures at which the high- and low-temperature approximations used in 1 and 2 cease to hold. Use as a criterion

a difference of 5% between the conductivity as extrapolated and the measured conductivity.

5. Choose a convenient temperature about halfway between the two temperatures found in 4. Calculate the conductivity at that temperature and compare with the measured value. Comment on any difference found.

A2.4 THE HALL EFFECT

A2.4.1 *General Arrangements*

When a magnetic field is applied in a direction perpendicular to the current flow in a material, a voltage is developed perpendicular to both the current and the magnetic field, as shown in Fig. A4. We can express this effect in terms of a *Hall coefficient R*.

$$\mathbf{E} = \rho \mathbf{J} - R\mathbf{J} \times \mathbf{B}$$

where the first term on the right represents the ordinary Ohm's law relation between current and electric field and the second term represents the Hall effect. For a magnetic field in the z-direction and current in the x-direction, we have

$$R = \frac{E_y}{J_x B} = -\frac{V_y/a}{IB/ab} = -\frac{bV_y}{IB}$$

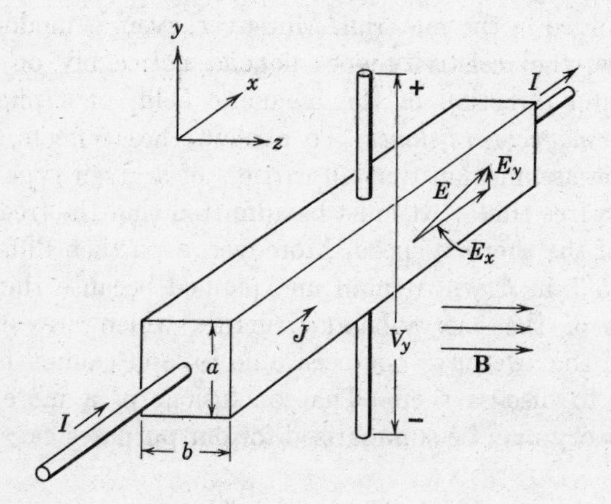

Fig. A4. Schematic arrangement for Hall effect.

The simple model of current flow developed in Sec. 1.5.6 gives the following relationship for the Hall coefficient in an extrinsic semiconductor:

$$n\text{-type, } R = -\frac{1}{nq}$$

$$p\text{-type, } R = \frac{1}{pq}$$

In this model, both ρ and R are independent of the magnetic field, regardless of its size. The sign of the Hall effect provides further evidence for the existence of two types of charge carriers.

The simple model of current flow may be extended to consider the Hall effect when both electrons and holes are present; the details are left as a problem (P1.19). The result for the Hall coefficient is:

$$R = \frac{1}{q}\frac{(p\mu_h^2 - n\mu_e^2)}{(p\mu_h + n\mu_e)^2}$$

if the magnetic field is not too large.

The simple quantitative results given above are in only fair agreement with experiment. In particular, the Hall coefficient R does depend on the strength of the magnetic field at larger fields, and at small fields its value differs from those quoted above by factors that depend on the specific kinds of carrier scattering processes involved in the material. Moreover, even at moderate magnetic fields, the resistivity does depend noticeably on both the strength and direction of the magnetic field, in a phenomenon known as *magneto-resistance.* To explain these effects, we must give up the assumption that all carriers of a given type have the same mean free time τ. It must be admitted that the free time is a function of the thermal speed. Moreover, even then differences as large as 25% in R will remain unexplained because they depend on details of the energy-band structure which are sufficiently specialized that we have not been able to, and cannot here, take the space to discuss them. The conclusions of a more comprehensive theory may be summarized for our purposes as:

1. For n-type germanium, $R = -\dfrac{0.93}{nq}$

2. For p-type germanium, $R = \dfrac{1.4}{pq}$

3. The magneto-resistance depends on B^2 at low and moderate fields, and is of the order of $\dfrac{\rho(B) - \rho(0)}{\rho(0)} = \mu^2 B^2$.

It should be emphasized that the additional features of the band structure responsible for the alterations of the Hall coefficient just mentioned do not have any major effects on the conduction properties in the absence of a magnetic field, except to change the numerical values of the mobility as computed from the theory of the scattering of charge carriers.

For the Hall effect, one experimental arrangement and the corresponding experimentally measured variables are shown in Fig. A5*a*. Two experimental difficulties arise in trying to achieve this arrangement and make the measurements.

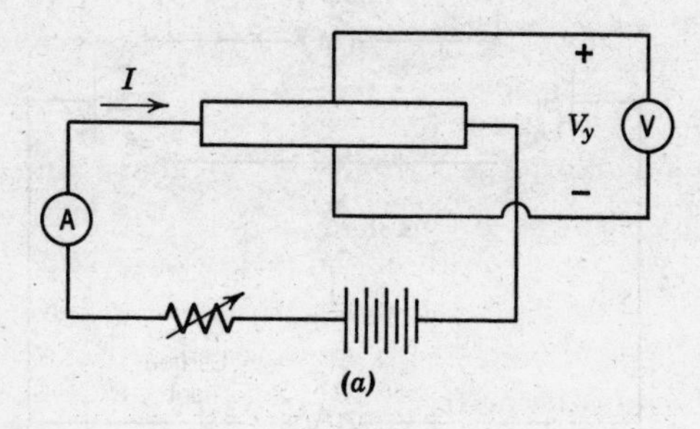

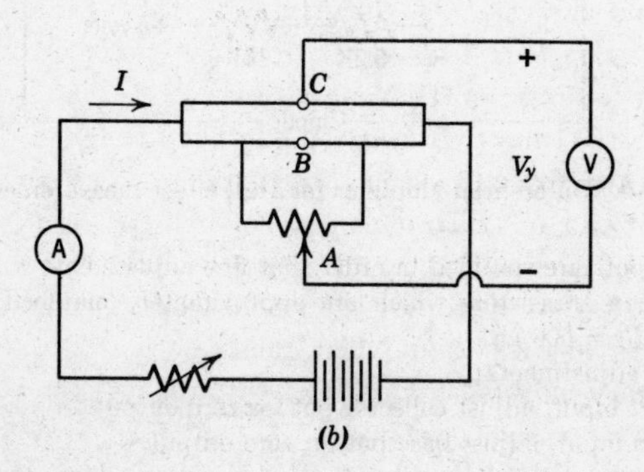

Fig. A5. Experimental arrangements for Hall-effect determination.

So that V_y will measure only the Hall effect, the two contacts must be aligned accurately opposite to each other. If there is some misalignment, V_y will include a component arising from the potential drop along the bar. To avoid this problem, we achieve the alignment electrically as shown in Fig. A5*b*. By varying the position of the moving contact on the potentiometer, we can make the potential at point A equal to that at point B. This is done by adjusting the potentiometer until V_y is zero in zero magnetic field. When a magnetic field is applied, the potential between A and C is the same as that between B and C as desired.

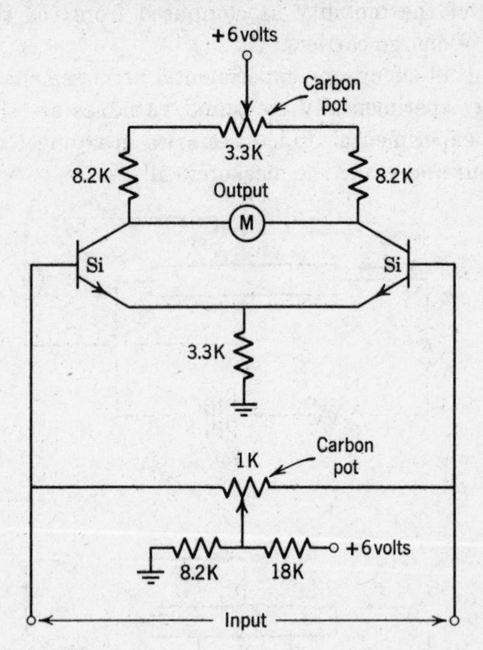

Fig. A6. Differential amplifier for Hall effect measurement. Notes:

1. Carbon pots are specified in order that fine adjustments will be simplified.
2. Use silicon transistors which are approximately matched, i.e., β within 10% at $I_C = 500\ \mu a$.
3. Balance adjustments:
   (*a*) Short input, adjust collector pot for zero output.
   (*b*) Open input, adjust base pot for zero output.
   (*c*) Repeat (*a*) and (*b*) until complete balance is achieved.
4. Check amplifier for drift; it may be necessary to protect amplifier from room temperature variations by covering the circuit.

The second problem is the small size of the Hall voltage (of the order of a few millivolts). We could increase the Hall voltage by increasing the current, I, or by increasing B. Both of these methods introduce experimental problems; e.g., sample heating and the eventual nonlinearity of the Hall voltage as a function of B. In Fig. A6 we show a sensitive measuring circuit suitable for use in this experiment. The operation of this circuit is discussed in Section 2.4.3.

A2.4.2 *Sample Mounting*

A simple method of mounting a sample with the required contacts so that it may be placed between the pole faces of a magnet is described below and shown in Fig. A7.

A wooden clothespin is prepared by removing the spring (which would be attracted to the magnet), and replacing the spring with a pencil or dowel and a rubber band, as shown in Fig. A7*a*. The jaws of the clothespin which will grip the sample should be sanded flat and parallel with coarse sandpaper. Take three pieces of fine tinned wire (single strands of ordinary stranded hook-up wire are suitable) about 4 in. long. Insert these in a second clothespin so that about $\frac{1}{2}$ in. of each of the three wires projects from the end of the clothespin (Fig. A7*b*). Separate the ends of the wires to form a triangle, with one wire above and two below spaced on either side. Clean your sample in acetone. Set the clothespin with its projecting wires on the edge of the bench and insert your sample with soldered end leads between the wires so that the three wires are near the center of the bar and arranged as required for the experiment. Using your prepared clothespin, grip the sample and the three contacts firmly. The long axis of the clothespin should coincide approximately with the long axis of the sample. Remove the clothespin which supported the wires and check that the wires are firmly held against the sample in the desired configuration. Using electrical tape, tape the clothespin shut so that the assembly will not come apart inadvertently. Tape should also be used to secure the five lead wires to the clothespin. Check the resistance between all pairs of contacts to your sample; the resistance between any pair should be on the order of the resistance of the bar. Be sure that no wires are shorted together. Connections to the measuring circuit should be made by soldering, using leads which are

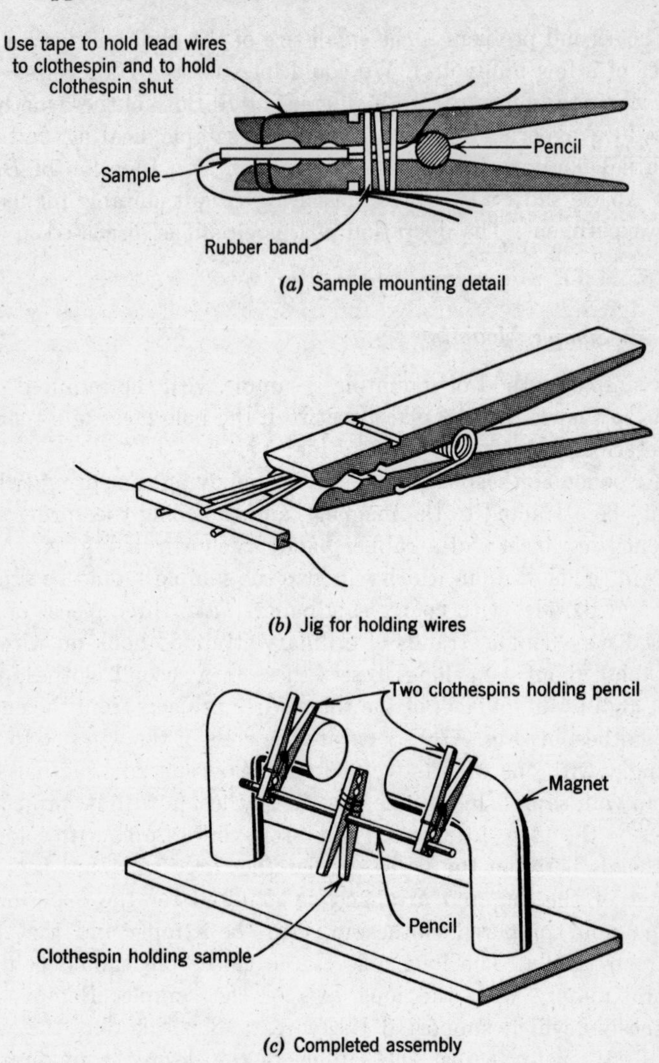

Fig. A7. Sample mounting for Hall effect determination.

long and flexible enough so that the sample may be inserted easily into the magnet. The sample may be held in the magnet by clipping two more clothespins to the pencil and using the magnetic force on the clothespin springs to hold them to the magnet, as shown in Fig. A7*c*.

A2.4.3 *Measuring Circuit and Technique*

Construct and test the difference amplifier shown in Fig. A6. Set up the measurement circuit shown in Fig. A8, taking particular care with the connections in the Hall voltage measuring circuit. Resistor R_1 is used to set the current through the sample. The differential amplifier is used as a null detector in this measurement. The variable voltage source and the voltage divider consisting of R_3, R_4, and R_5 are arranged so that the difference amplifier measures the difference between V_b and the Hall voltage. The voltage divider should be set carefully with a precision voltmeter so that V_b is 0.001 times the reading of V_2.

The measurement procedure is as follows. With the difference amplifier balanced and V_2 reading zero, adjust R_2 so that the meter in the amplifier (V_1) reads zero when the sample is not in a magnetic field. Place the sample in the magnetic field and adjust the variable voltage supply V_a until the amplifier meter again reads zero. Then V_2 reads 1000 times the Hall voltage. *Be sure* you understand the function of this circuit before proceeding.

Measure the Hall voltage as a function of sample current and of magnetic field (using several different permanent magnets of known field strength) for the *n*-type and *p*-type samples. Does the Hall voltage behave as expected when you reverse the sample current and when you reverse the magnetic field? Observe qualitatively the effect of rotating the sample about its long axis in the magnetic field.

After obtaining all the necessary data on the Hall effect, you may measure magneto-resistance if time permits. Since the magneto-resistance is larger in *p*-type than in *n*-type material, it would be wise to start with the *p*-type sample. The change in resistance in a field of 2000 gauss is about 0.5%, and in a field of 4000 gauss the change is about 2%. This means you will have to devise a circuit for measuring small changes in resistance; use your Hall effect measuring circuit as a guide.

A2.4.4 *Summary of experimental procedure*

1. Measure sample dimensions.
2. Prepare sample for Hall measurement.
3. Measure Hall voltage and sample current in the available magnetic fields.

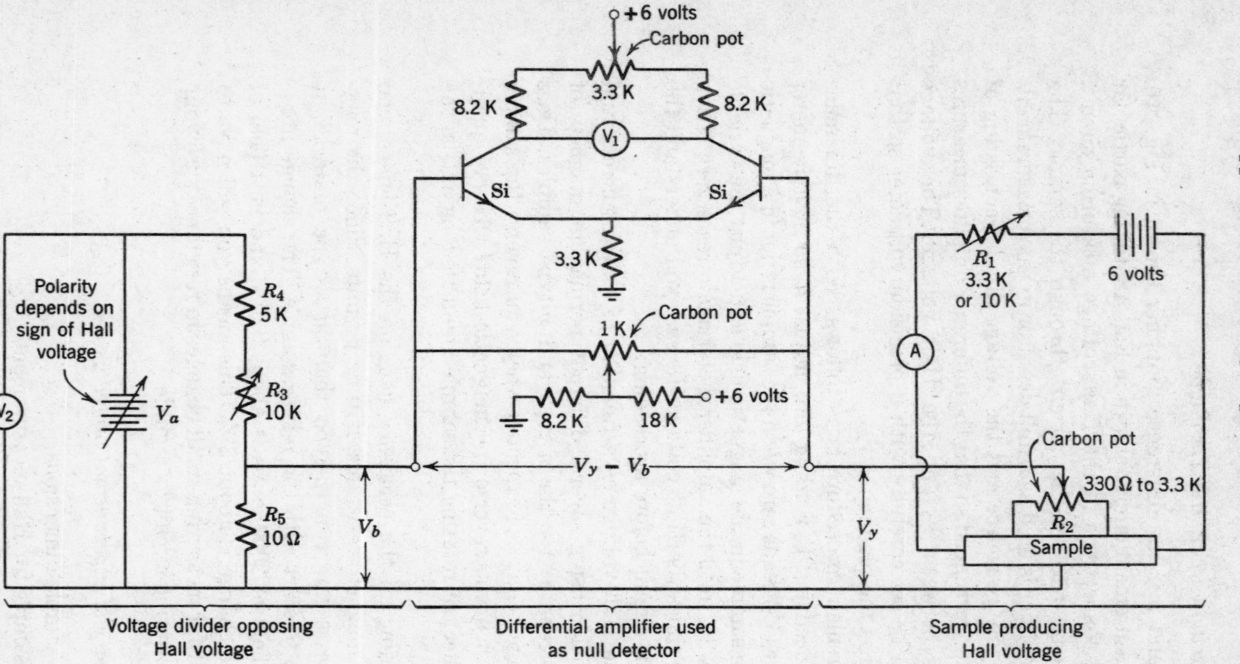

Fig. A8. Measurement circuit for Hall effect. Procedure:

1. Balance differential amplifier (see Fig. A6).
2. Set voltage divider, using R_3 and precision meters so that $V_b = V_a/1000$.
3. Adjust R_1 for desired current.
4. With no magnetic field and $V_a = 0$, adjust R_2 until $V_1 = 0$.
5. In the magnetic field, adjust V_a until $V_1 = 0$ again.
6. Then $V_y = V_a/1000$.

A2.4.5 *Analysis of data*

1. Show that the Hall voltage has the sign expected for the majority-carrier type of each sample.

2. From your data on Hall voltage, sample current, magnetic field and sample dimensions, calculate the Hall coefficient, and hence the majority carrier concentrations in each sample. Compare with the result of your conductivity measurements.

3. The Hall mobility is defined as R/ρ. Using the numerical factors for R on pp. 206–207, your measured value of Hall mobility may be compared with the values given in Table 1.2.

PRODUCTION OF ELECTRON-HOLE PAIRS AND THE ENERGY GAP†

A3.0 INTRODUCTION

One important aspect of our model of semiconductors is the fact that we can vary the concentrations of minority and majority carriers by external means. In the analysis of diodes and transistors, this variation is produced by injection across a *pn* junction. Experimentally, it is easier to obtain direct evidence of carrier pair production by other means. One of the simplest methods, and the one we use, is the creation of electron-hole pairs by photons. By observing the relation between photon energy and pair production, we are able to measure the size of the energy gap involved in the production of a hole-electron pair (see Secs. 1.6.1 and 2.2.2 and Problem P2.4). This we do for both Ge and Si.

A3.1 OPTICAL ABSORPTION AND PHOTOCONDUCTIVITY

If we have an experiment in which photons of a single energy (monochromatic light) are falling on a well-polished semiconductor, various things happen. A fraction of the incident photons is reflected because the index of refraction of the semiconductor will, in general, be different from that of air. Of those photons which

† A different experiment which shows the energy gap is performed in the SEEC Film referred to on p. 59.

enter the semiconductor, some may produce electron-hole pairs and some may be absorbed by other energy-consuming processes within the solid. The remaining photons may be reflected at the back surface of the solid or may be transmitted. The relative importance of these various phenomena will depend on the energy of the photons and on the particular semiconductor involved. We wish to focus attention on the production of hole-electron pairs, and in particular on the energy required to produce pairs.

Two types of measurement are suggested by the discussion above. We could measure the fraction of incident photons which are transmitted, and thus infer something of the processes going on inside the semiconductor. Or we could measure a property of the semiconductor that depends on the concentration of electrons and holes; e.g., the conductivity. We use both methods in this experiment.

Specifically, if we have a means of selecting various incident photon energies, we may look at both the fraction of transmitted photons and the photoconductivity of the semiconductor as functions of photon energy. From our model of semiconductors, we would expect that for photon energies less than the energy gap no pairs would be produced, and the results of the experiments would depend on other processes. When the photon energy is increased to be equal to or greater than the energy gap, we would expect to see a sharp increase in the photoconductivity and a sharp decrease in the fraction of transmitted photons. From the photon energy at which these effects occur, we may infer the size of the energy gap. In principle, we can also analyze the detailed dependence of this absorption process on photon energy and obtain correspondingly more information about the details of the interaction. However, this analysis is not simple and requires careful experimental technique, along with information (and assumptions) about the incident light, the measuring apparatus, and the semiconductor.

A3.2 THE DIFFRACTION-GRATING SPECTROMETER

From the discussion above, it is evident that we need a way of selecting monochromatic photons (more precisely, a group of photons with only a small range of energies), and a means of selecting the energy of these photons over a suitable range. To do this, we use the simple spectrometer shown in Fig. A9*a*.†

Such a spectrometer consists of: a source of light; a dispersing element, which causes light of different wavelengths (photons of different energies) to travel along different paths in space; a slit or collimator to select only those wavelengths which are following a particular path in space; and finally, a detector of radiation.

As a source of light, we use an incandescent bulb such as an automobile headlight bulb.‡ We also require that the light rays be parallel; so we need to use a lens with the bulb filament at the focal point.

The dispersing element is a plastic replica diffraction grating. The grating consists of a large number of closely spaced lines. Light of different wavelengths is diffracted by the grating at different angles. By varying the angle of incidence of the light and looking at different angles of diffraction, we can choose groups of photons with different energies. We discuss this more quantitatively in the section on calibration of the spectrometer.

The collimator used here consists of a number of soda straws packed into a cardboard tube. Light rays which are parallel to the axis of the tube pass through the collimator; those at an angle to the axis are absorbed by the walls of the straws. A lens following the collimator focuses the light (through a thin silicon sheet) onto one-half of a germanium sample. In our system, the germanium sample acts as the detector of radiation; we observe the change in conductivity arising from the incident light.

† It is not possible to use the same Ge samples as in the previous experiments because of problems with surface preparation and the achieving of adequately small thickness. Sets of suitable samples of both germanium "dendrite" and silicon "web" are available from Educational Services, Inc., 164 Main St., Watertown, Mass. Detailed instructions for construction of the spectrometer are also available from the same source. The germanium dendrite and silicon web employed here were originally developed by Westinghouse Research Laboratories, Pittsburgh, Pa.

‡ A suitable bulb is one like Westinghouse No. 1134, 6-8 volt, 4 amp, 32 CP. Equivalent ones with different bases are Westinghouse No. 1323 and 1133 (the double-filament type like No. 1000 is also satisfactory). An unfrosted 150-watt tungsten lamp, with a linear helical filament, will also work, but needs a larger condensing lens. Sunlight gives the largest responses, but is less convenient to use.

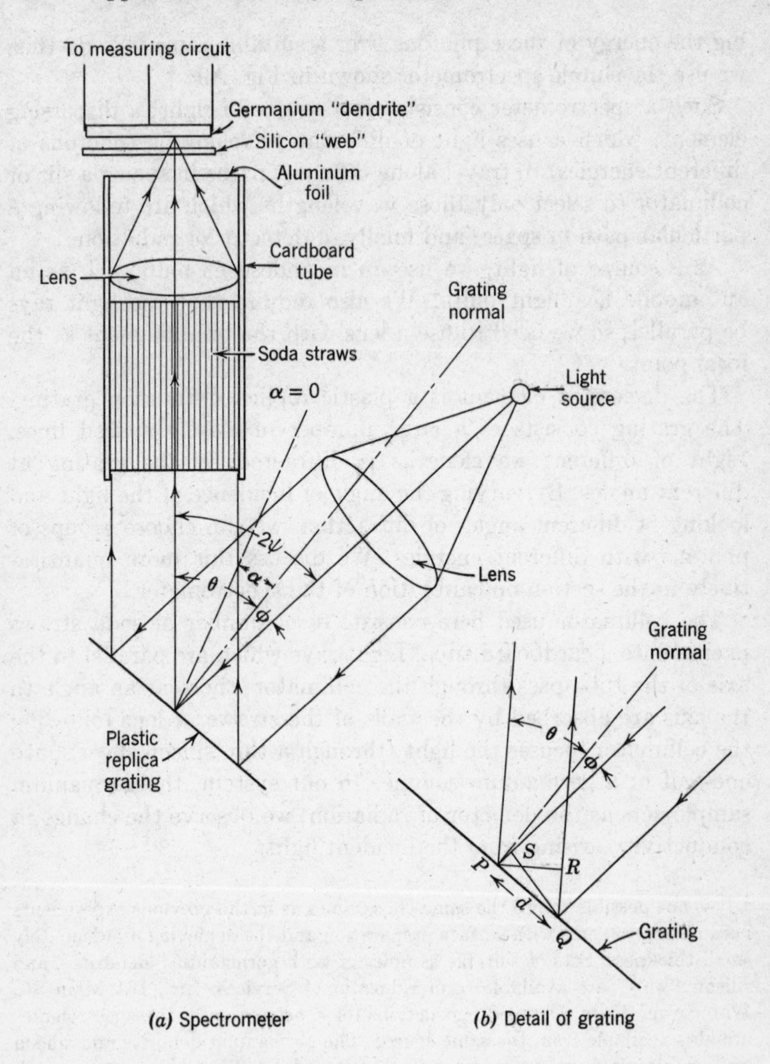

Fig. A9. Spectrometer geometry for energy-gap determination.

A3.2.1 *Calibration of Spectrometer*

We assume that you are familiar with the basic idea of a diffraction grating. A simplified expanded picture of the grating is shown in Fig. A9*b*. The condition for constructive interference of the light rays is that the path difference (PS-QR) must be an integral number of wavelengths. In terms of the figure,

$$\sin \phi - \sin \theta = \frac{n\lambda}{d}$$

where d is the grating line spacing and n is any integer, positive, negative, or zero; n is called the *order* of the diffracted spectrum.

Let us first estimate the order of magnitude of the wavelengths with which we will be dealing. The energy gap in silicon is about 1 ev and in germanium it is about 0.7 ev. The energy in electron volts of a photon of wavelength λ in Angstroms is

$$E(\text{ev}) = \frac{12395}{\lambda(\text{Å})}$$

so that λ_{Ge} is about 18,000 Å or 1.8 microns, and λ_{Si} is about 12,000 Å or 1.2 microns. Both wavelengths are in the infrared.

For our spectrometer, the sum of the angles ϕ and θ is fixed during the experiment and the wavelength which passes through the collimator is determined by rotating the grating. Let us define

$$2\psi = \phi + \theta; \qquad 2\alpha = \phi - \theta$$

so that

$$\phi = \psi + \alpha; \qquad \theta = \psi - \alpha$$

and the relation between wavelength and angle becomes

$$2 \cos \psi \sin \alpha = \frac{n\lambda}{d}$$

Because ψ is fixed during rotation of the grating, we need to find the value of $(2d \cos \psi)/n$ to express λ in terms of α. One way to do this is to measure the angle 2ψ between the light source and the collimator and use the grating spacing d given by the manufacturer, 13,400 lines per inch.

To find the position $\alpha = 0$, we note that the zero-order spectrum ($n = 0$) occurs at $\alpha = 0$ and that the grating equation is satisfied for all wavelengths simultaneously. Hence the zero-order diffraction will consist of white light.

As an experimental check on the grating arrangement, start from $\alpha = 0$, rotate the grating, and observe the angles at which green light passes through the collimator. Do this for both directions of rotation. The light passing through the collimator may be observed easily by using a piece of thin, white paper. The wavelength of green light is about 5500 Å. From this fact and these measurements, you can determine an experimental value of the calibration constant and compare it with the value calculated above.

A3.3 DETERMINING GERMANIUM ENERGY GAP

In the experimental calibration just described, we were able to obtain calibration points useful for infrared radiation by using only visible light. This is possible because n and λ enter the grating equation only through the product, $n\lambda$. In other words, for the same grating position we have light of several wavelengths passing through the collimator. In particular, if we set the grating so that the first-order wavelength corresponds to E_g(Ge), $\lambda_1 = 18{,}000$ Å, light of wavelengths $\lambda_2 = 9000$ Å, $\lambda_3 = 6000$ Å, $\lambda_4 = 4500$ Å corresponding to $n = 2, 3, 4$, will also fall on the germanium. (There is, however, very little intensity of light of wavelength less than 4500 Å because of the emission characteristics of the source and the absorption characteristics of glass). The shorter wavelengths (larger photon energies) will produce photoconductivity in the germanium obscuring the cut-off we are looking for near 18,000 Å. In fact, for our light source the intensity at the 9000 Å and 6000 Å wavelengths is probably much greater than that at 18,000 Å.

To eliminate the unwanted wavelengths of multiple orders, we would like to have a filter which eliminates wavelengths below about 10,000 Å. For this purpose, we use a semiconductor whose energy gap is somewhat less than twice the energy gap of germanium; e.g., silicon. For moderate thicknesses, the absorption accompanying the production of electron-hole pairs is very strong, so that silicon is a very good filter for wavelengths less than about 12,000 Å. (For longer wavelengths, silicon does not absorb the radiation; however, its index of refraction is such that it reflects about 40% of the light and transmits about 60%.) Using silicon as a filter also allows us to measure the energy gap of silicon, inasmuch as no photoconductivity of the germanium will be observed for wavelengths less than that corresponding to E_g(Si).

To summarize, we have drawn in Fig. A10*a* some curves which show the relative intensity of radiation from the source, the rela-

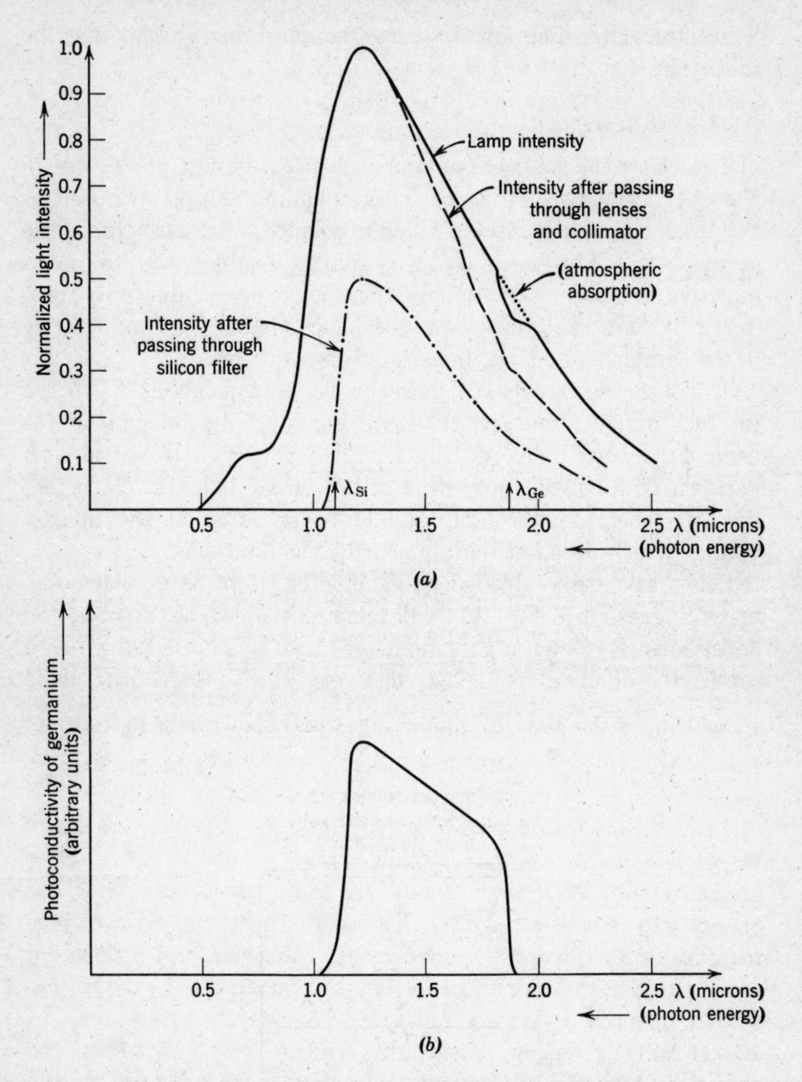

Fig. A10. Optical properties of the spectrometer of Fig. A9.

tive intensity passing through the spectrometer, the intensity of radiation passing through the silicon filter, and in Fig. A10*b* the photoconductivity produced in the germanium. Note that the decrease in photoconductivity near 18,000 Å occurs at a wavelength where the emission from the bulb is falling off sharply, and hence the determination of the germanium energy gap may be uncertain.

A3.3.1 *Measurements*

The electrical connections and measuring circuit are shown in Fig. A11. The two halves of the germanium sample are used as two arms in a bridge circuit which is balanced with no light on the germanium.† This type of measurement enables you to use a sensitive detector to read only the change in conductivity produced by light, and also *tends* to balance changes in conductivity in the sample caused by changes in temperature.

The observed signals are, however, very small (about 0.5 to 1μa for the configuration given). Switching the lamp on and off (or using an infrared opaque shutter) will help you to distinguish between photo-signals and drift of the bridge balance. Nevertheless, the spectrometer lamp should be operated at the highest voltage which does not unduly shorten the lamp life.

Using a sensitive bridge galvanometer in place of the 50-μa meter suggested in Fig. A11 will increase the size of the observed deflections. But such a galvanometer must be shunted down to a sensitivity of about 0.1μa per mm, otherwise bridge drift alone

† The theory of this arrangement forms the basis of problem P4.13.

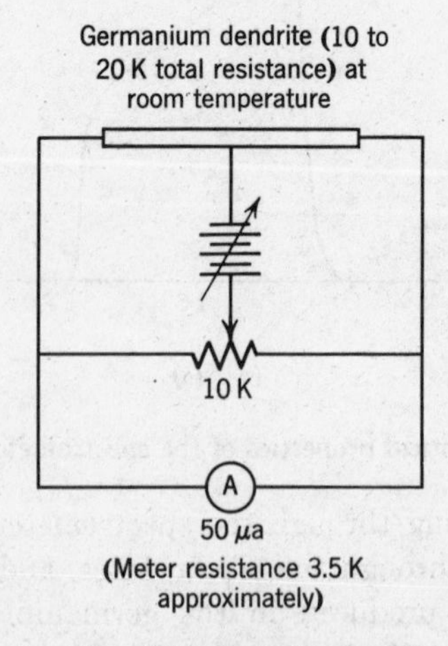

Fig. A11. Detector circuit.

may drive it off scale. Observe the bridge impedance level when choosing galvanometer and shunt.

A3.4 SUGGESTED PROCEDURE AND ANALYSIS OF DATA

1. Show that if the conductivity changes are a small fraction of the dark conductivity, the meter deflection is proportional to the change in conductivity.
2. Calibrate your spectrometer as described above. Is the approximation $\sin \alpha \simeq \alpha$ a valid one to use for this experiment?
3. In a bridge circuit that uses ideal resistors, the sensitivity increases linearly with source voltage. Using a white light (zero-order position of the grating), determine experimentally the dependence of sensitivity on source voltage. Discuss any deviation from the ideal case.
4. Using a source voltage which gives best sensitivity, measure the relative photoconductivity as a function of wavelength and determine the energy gaps of germanium and silicon. Estimate the accuracy of your determination and compare with published values (see Table 1.1).
5. Considering the difference between light intensity described in watts and in terms of number of photons per second, explain why the shape of Fig. A10*b* differs from that of the dash-dot curve in Fig. A10*a*.
6. What considerations determine the choice of ψ in the construction of the spectrometer for this experiment (see Fig. A9)?

DYNAMICS OF EXCESS CARRIERS†

A4.0 INTRODUCTION

Excess carriers, both holes and conduction electrons, may be created in a semiconductor by a variety of methods; e.g., by shining light on the sample, by appropriately biasing a metal point contact, or by biasing a *pn* junction. The principal dynamic features of

† A complete form of the classic experiment on this topic is carried out in the SEEC film referred to on p. 181.

these excess carriers are that they *drift* in an electric field, they *diffuse* away from regions of high concentration, and they *disappear by recombination*. These three effects are represented quantitatively by the three terms on the right in the diffusion equation (see Chapter 4):

$$\frac{\partial p}{\partial t} = -\mu_h E \underset{\text{(drift)}}{\frac{\partial p}{\partial x}} + D_h \underset{\text{(diffusion)}}{\frac{\partial^2 p}{\partial x^2}} - \underset{\text{(recombination)}}{\frac{p - p_o}{\tau_h}}$$

The parameters used to describe the effects are mobility μ, diffusion coefficient D, and lifetime τ. In this experiment, we try to measure the drift mobility μ and observe qualitatively the effects of diffusion and recombination. We perform the experiment on n-type germanium, and observe the properties of the excess *minority* carriers, the holes.

A4.1 MEASUREMENT OF DRIFT MOBILITY

A severe limit is put on the experimental arrangement by the existence of recombination. The excess carriers which we inject last, on the average, for a lifetime. In your samples this time is of the order of 10 to 20 microseconds. Thus any experiment on the dynamics of these carriers must take place in about this time interval. It will make the experiment easier and more meaningful if we take pains to make the lifetime as long as possible. This can be done by careful sample preparation.

The basic idea of the measurement of drift mobility is straightforward. We create a local distribution of excess carriers near a point; then in an electric field, we measure the time it takes for the distribution to drift a known distance. We thus know the velocity of the carriers, and by measuring the electric field we can calculate the mobility from the relation $v = \mu E$.

If we calculate the order of magnitude of the quantities involved, some of the experimental difficulties will become evident. To make a reasonably accurate measurement of the distance traveled, it should be at least a few millimeters. The time for the carriers to drift this distance will have to be about 20 microseconds or less; otherwise most of the carriers will recombine before we can detect them. Thus the drift velocity must be 2×10^4 cm/sec or greater.

From our Hall measurements, we estimate the drift mobility of holes to be approximately 2×10^3 cm^2/volt sec, so we need to use an electric field of 10 volt/cm or more. For a sample 1 cm long, having a resistance of 200 ohms, this means we have to dissipate a power of half a watt or more. This would heat the bar significantly and, as we found in the conductivity measurement, the mobility is a function of temperature. In fact, since the mobility decreases at high temperatures, the calculation above may have been optimistic. At high temperatures, too, the presence of both electrons and holes in equilibrium may mean that the sample is no longer extrinsic and this complicates considerably the interpretation of the experiment.

One way around this self-heating problem is to turn the electric field off for most of the time, and only turn it on when we are actually observing the excess carrier distribution. In order to display the effects, we will use a repetitive square pulse of duration about 50 microseconds, repeated every 10 milliseconds. The rate of temperature rise of the sample is governed by its specific heat and is slow enough so that we need to consider only the average power in the sample for the conditions given above. The average power is just 5×10^{-3} of the power previously calculated for the dc case, and the temperature rise is probably not troublesome. A transistor circuit for producing the pulses is described in Sec. A4.2.

To inject and observe the excess carriers, we use metal point contacts which have rectifying characteristics. The basic arrangement is shown in Fig. A12*a*. One point contact, the *emitter*, is forward-biased as shown; under these conditions, part of the current which flows into the sample consists of excess holes. While the electric field is off, these excess holes do not drift, but only diffuse and recombine. This gives rise to a distribution of holes around the emitter point of the form,

$$p - p_o = p' = ce^{-|x|/\sqrt{D\tau}}$$

When the electric field is turned on, the emitter immediately becomes reverse-biased and the distribution of excess holes drifts down the bar toward the collector. Because of diffusion and recombination, the distribution changes shape and size, but the maximum travels with the drift velocity. Thus the time at which

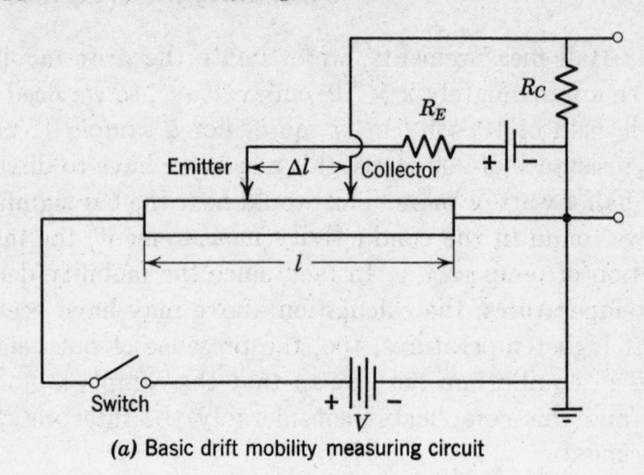

(a) Basic drift mobility measuring circuit

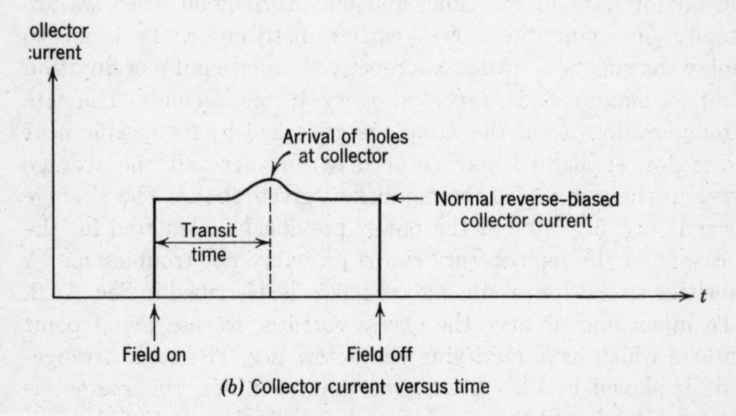

(b) Collector current versus time

Fig. A12. Schematic arrangement for measurement of drift mobility.

the maximum arrives at the collector is given by $t = \Delta l/\mu E$. The presence of excess holes under the collector is detected by observing the current through R_c. The collector is a reverse-biased point contact, the reverse bias being supplied by the source that applies the field to the germanium. When the field is on, very little current ordinarily flows in the collector, because the collector point will be at a lower potential than the germanium immediately under it. The small reverse current that does flow in the collector, however, depends on the concentration of holes in the neighborhood of the point; so it increases when the excess holes arrive at the collector. The current in the collector circuit is shown in Fig. A12*b*.

The experiment may be performed for various values of the electric field and thus the hole pulse may be observed at different times with the same emitter-to-collector spacing. In this way, we may observe qualitatively how the pulse widens with time because of diffusion, and how the pulse area decreases with time as a result of recombination.

A4.2 PULSE GENERATOR

The circuit for the pulse generator is shown in Fig. A13*a*. Q_1 and Q_2 are connected in an astable multivibrator circuit with a short duty cycle. The off-time of Q_2 is determined by R_1C_1 and is adjustable from 10 to 110 microseconds. The off-time of Q_1 is determined by R_2C_2 and is set at about 9 milliseconds. Q_3 is used to provide fast rise and fall times of the pulse. Q_4 is an overdriven common emitter amplifier whose collector voltage swing may be limited by R_7 to adjust the output pulse height. Q_5 is an emitter follower to provide the power to the bar.

Construct the circuit as shown and test with a resistive load in place of the sample. Before connecting the 45-volt battery, connect the 6-volt battery and check the operation of the multivibrator. The waveform at the collectors of Q_2 and Q_3 should be as shown in Fig. A13*b*. Unless Q_2 is saturated for at least 99% of the time as shown, the average power dissipated in the load, Q_4, Q_5, and R_7 may be excessive and result in their destruction. After this has been checked, connect the 45-volt battery and observe the output waveform. It should be as shown in Fig. A13*c* with the pulse height adjustable from zero to 39 volts by varying R_7 (the highest voltage requires removing R_7 from the circuit completely).

A4.3 SAMPLE PREPARATION

For this experiment, the preparation of the sample and contacts is critical. The lifetime of excess carriers depends not only on the perfection and purity of the sample but also on the state of the surface. Since we have no control over the internal state of the sample, we concentrate on producing a good surface. A good surface also makes the problem of producing suitable point contacts easier. Be sure to use the *n-type* bar because it is quite difficult to produce suitable point contacts on *p*-type material.

Remove the lead wires from your sample, leaving the end surfaces tinned with solder. Remove any solder on the long surfaces by rubbing lightly with fine emery paper. The long surfaces are then polished to a mirror finish with Linde A lapping powder on a glass

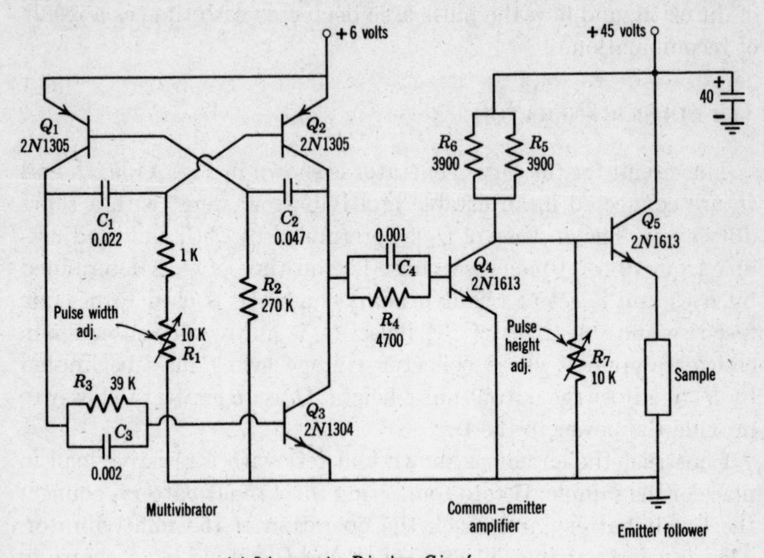

(a) Pulse generator: R in ohms, C in μf

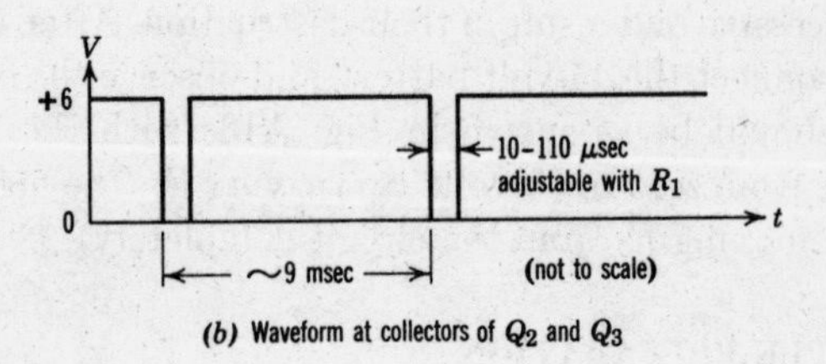

(b) Waveform at collectors of Q_2 and Q_3

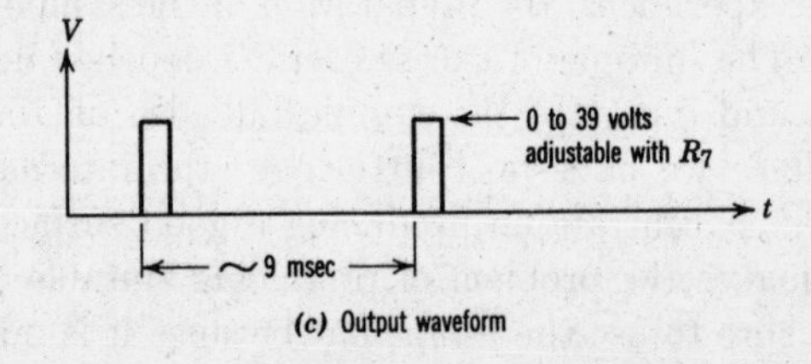

(c) Output waveform

Fig. A13. Pulse generator and its operation.

plate. Care in the polishing operation leads to much better results later. After polishing, the bar should be washed in water to remove any lapping powder. The lead wires (#26 stranded wire) are then soldered to the ends, taking care not to get solder or flux on the polished surfaces. The sample should then be washed in acetone and in water.

Just before mounting the sample for the experiment, it should be etched in boiling 3% hydrogen peroxide for three minutes. Rinse the bar in clean water and allow to dry. Do not touch the prepared surfaces—handle the sample by the lead wires.

The sample must be mounted so that the point contacts may be applied in a mechanically stable manner. A method for doing this is shown in Fig. A14. The wires to be used for point contacts are

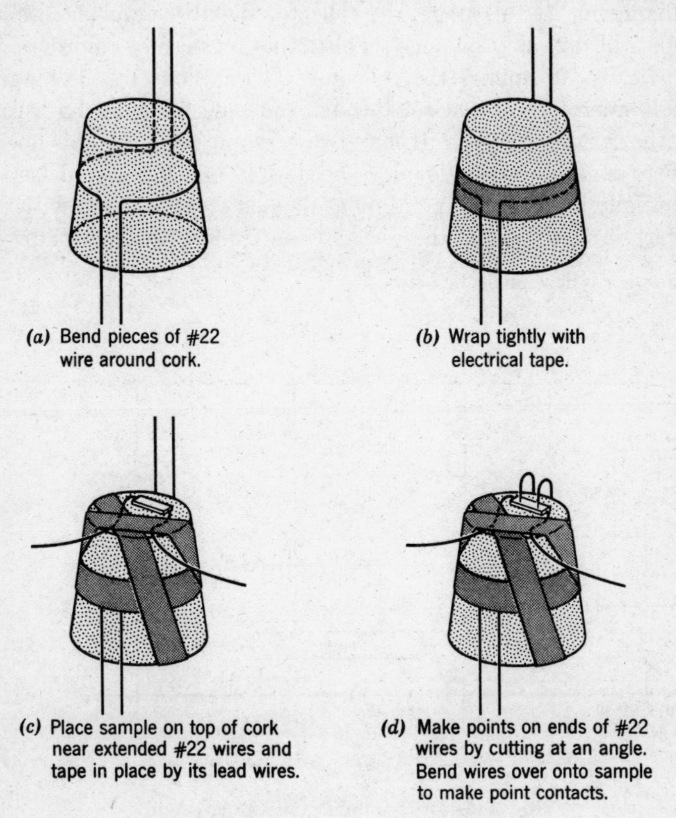

Fig. A14. Sample mounting instructions.

#22 bare tinned copper wire.† These are taped rigidly to a 1 in. cork as shown. The sample is then held rigidly on the top of the cork by taping the lead wires as shown. The points are made by cutting the wires at an angle and then bending them to make contact with the sample. The points should be 3 to 5 mm apart and near the center of the bar. The points should press on the bar with some force so that they will not be easily moved when handling the assembly. The rectifying nature of the points may be checked on a transistor curve tracer, an ordinary oscilloscope, or very roughly with an ohmmeter.

The sample is then connected in the measuring circuit as shown in Fig. A15. Set the pulse generator for pulses of 40 μsec duration and 25 volts amplitude. Adjust the emitter current to about 1 milliampere. At this stage, the collector signal seen on the oscilloscope will probably be noisy. This situation may be improved by electrically "forming" the collector. This is done by discharging the 40-microfarad capacitor through the collector point by throwing the switch as shown. It may be necessary to repeat this operation to obtain a good signal. (The desired effect involved here is essentially to place the metal point in good stable contact with the surface, and to obtain the rectification from a semiconductor *pn*

† Phosphor bronze might be better.

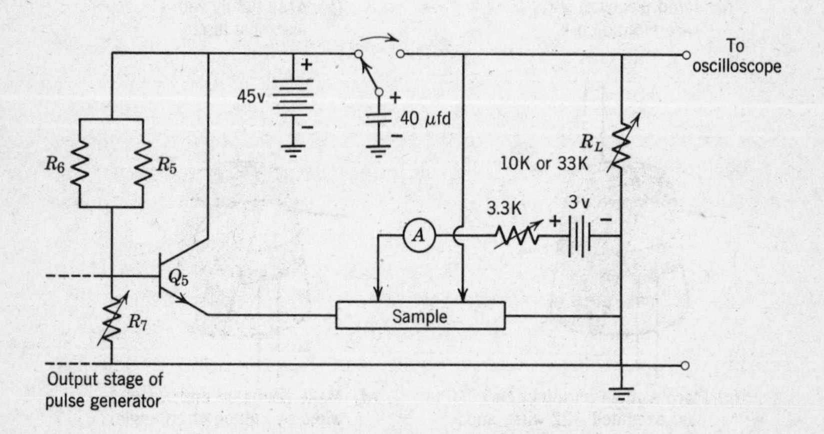

Fig. A15. Drift mobility measuring circuit.

junction below the surface. Overforming will, however, effectively weld the contact, making it ohmic and useless as a collector. If that happens, the point must be moved to a different spot, and re-formed.) Be sure the switch is in its normal position before proceeding.

A4.4 MEASUREMENT PROCEDURE

You may see a spike or a rounding-off of the leading edge of the collector signal. This probably arises from the capacitance of the collector point and may be altered by changing R_L and hence the effective bias of the collector. Adjust R_L to give a square leading edge—this adjustment will have to be repeated whenever the electric field in the sample is changed.

The shape of the collected hole pulse should be nearly symmetrical. An asymmetrical pulse is usually an indication of over-injection at the emitter. For this reason, the emitter current should be kept as low as possible when making measurements.

With no emitter current, the collector waveform should be square. If adjustment of R_L does not give a flat signal, the trouble may arise from poor end contacts which are injecting excess carriers into the sample and thus producing a nonuniform electric field. To check this, connect a resistor in series with the bar and observe the waveform across this resistor; i.e., the current in the bar. It should have the same shape as the voltage waveform; if it does not, the difficulty is probably with the end contacts. The remedy is to remove the end contacts and resolder them.

Once a satisfactory signal is obtained at the collector, proceed to measure the transit time of the holes as a function of the voltage across the bar. The transit time is the time interval from the moment the field is turned on, until the maximum of the hole pulse arrives at the collector. The voltage across the bar may also be measured with the oscilloscope by moving the scope probe from the collector to the end of the bar. (A dual input scope makes this more convenient.) Take measurements over as wide a range of voltage as possible. You can check whether the bar is heating appreciably by varying the length of time the electric field is on. If an increase in average power does not shift the time of arrival of the hole pulse, you may conclude that the mobility is not changed because of heating. In addition to the values of transit time and voltage, you will need to know the length of the bar and the separation between the contacts.

A4.5 SUGGESTED ANALYSIS OF DATA

1. According to the relation between transit time and field, a plot of V versus $1/t$ or a plot of t versus $1/V$ should yield a straight line. Plot your data both ways and comment on any discrepancies observed. Decide which plot to use and, from the slope of the line and the sample dimensions, calculate the mobility. Compare with values given in Table 1.2. Comment on any differences observed.
2. State your qualitative observations of diffusion and recombination, and give an estimate of the lifetime.
3. Discuss carefully all the conditions you can think of under which, in this experiment, the observed peak of the minority-carrier pulse at the collector would *not* occur at a time $\Delta l/\mu_h E$ after the field is applied.

MANUFACTURE OF AN "ALLOY" DIODE†

A5.0 INTRODUCTION

The simplest theory of a *pn* junction predicts a dc volt-ampere characteristic of the form:

$$I = I_o(e^{qV/kT} - 1) \tag{A1}$$

where I is the current flowing through the junction from the p-type to the n-type side, and V is the voltage by which the p-type side is positive with respect to the n-type side. In this simple theory, I_o is independent of V and has two possible interpretations:

(1) It is the "reverse saturation current" in the sense that

$$I \rightarrow -I_o \text{ when } V \ll -\frac{kT}{q} \; (\approx -26 \text{ mv at } 300°\text{K})$$

† This experiment can be done most profitably only after studying the *pn* junction. See, for example, Chapters 1 to 4 of PEM.

(2) It is the zero-voltage intercept of the forward $V - I$ characteristic extrapolated back from high currents.

$$\ln I \rightarrow \ln I_o + \frac{q}{kT} V \text{ when } I \gg I_o$$

In a real diode, a number of physical effects occur which are not included in the aforementioned simple theory. These effects become more or less important, depending on the geometrical structure and metallurgical method of fabrication of the diode. Some of the variations of structure and metallurgy cause major departures from the simple volt-ampere characteristic presented above. Others produce only small deviations.

In this experiment we wish to examine some of the physical factors that are important in making an "alloy" type of *pn*-junction diode, and to determine the relationship of these factors to the volt-ampere characteristic of the resulting structure. The experiment will be successful if, as a result of doing it, you gain some understanding of what must have been *left out* of any discussion of *pn*-junction diodes which focuses attention mainly on getting Eq. A1 for an answer. In this respect, it is very helpful (though not essential) to examine the results of *each step* in the fabrication process under a good microscope, if one is available. The more you can actually *see* in this experiment, the more you will learn from it.

A5.1 PREPARATION OF THE WAFER

A5.1.1 *Tinning*

As a first step in the diode-making procedure, we must provide for a contact to the wafer which will be of low resistance and act neither as an injector or collector of minority carriers. Such a contact is said to be "ohmic." The following procedure will establish the basis for an ohmic contact.

1. With fine emery paper, roughen slightly one surface of the *n*-type 1 to 5 ohm-cm Ge wafer (see p. 195). Then with a damp tissue remove any dust from the wafer surface.
2. Wet the roughened surface of the wafer with Polyflux (see p. 196).

3. Put some coreless solder† onto the small tip of the hot soldering iron and touch the iron and melted solder to the roughened side of the Ge wafer.

Care should be taken to insure that only the roughened side of the wafer is tinned. The entire roughened side, however, should be tinned to provide an ohmic contact of the largest possible area.

After tinning, in most cases, the solder will be in the form of a lump on the wafer. This lump should be sanded flat with emery paper so that the wafer may rest parallel to the hot plate surface during the subsequent alloying stage.

A5.1.2 *Polishing and Etching*

In order to get even a reasonably uniform geometry and depth of penetration of the junction we are about to make, the major surface damage produced by the original sawing of the wafer must be removed. This means we must polish and etch the top surface of the wafer. In addition, because this experiment has been planned for normal execution without the use of dangerous chemicals, the surface we start with is the one that ultimately lies adjacent to the junction we shall prepare. Thus three characteristics of the completed diode will probably be improved by careful surface preparation initially:

1. The "saturation current" will more nearly "saturate." A layer of disordered material, grease, or other conducting compound on the surface, near the place where the junction emerges, may act as a relatively low-resistance bridge across the junction. The electric field near the junction under reverse-bias conditions is very high (10^5 volts/cm or more). So, unless such a bridge is removed, it can easily dominate the reverse characteristic, giving it an appreciable slope.

2. The saturation current will be reduced. Surface damage (cracks, crevices, pits, etc.) increases greatly the effective surface area presented to carriers inside the material. All surfaces are, of course, places of discontinuous crystal structure, and accordingly

† The type specified for the conductivity experiment is suitable; but for the diode the percentage of lead is not critical if aluminum is used over the hot plate later, as shown in Fig. A16.

they furnish abnormal places for carriers to reside. Among such places are *recombination centers* (and, by detailed balancing, these are *generation centers*, too). A recombination-generation center is simply an intermediate stepping stone between the conduction and valence bands, which increases the probability that electrons can make the whole trip. Their presence therefore increases the thermal generation rate, and correspondingly the reverse saturation current.

3. The maximum tolerable reverse voltage will be raised. The bridging effect discussed under item 1 above becomes very severe at high fields and causes large increases in back current. This nevertheless happens more gradually with voltage than either avalanche or Zener breakdown,† and such bridging leads to what is called a "soft" breakdown, unless it is removed.

Polishing of the wafer is accomplished with Linde "A" lapping powder and a drop of water on a glass plate, or on metallurgical lapping paper glued to the plate. The untinned surface of the Ge wafer should be polished to a mirror finish. When polishing is completed, wash the sample well with water.‡ From this point on the wafer should be handled only with tweezers or the equivalent.

The Ge wafer is now etched by placing it in a beaker containing 6 to 10 ml of H_2O_2 USP 3%.§ Because of the mildness of the etch it is necessary to boil the wafer in the 3% solution for 2 to 3 min. Following the etch, the wafer is washed in water and allowed to dry completely.

A5.2 ALLOYING

The purpose of alloying is to form a *Ge-In molten solution which, upon cooling, will recrystallize into a region of heavily p-doped single-crystal Ge.*

† See PEM.

‡ Distilled or demineralized water does not seem to be *essential* for the washing operations in this experiment, unless the local water condition is particularly poor; but improved reverse characteristics generally result from its use.

§ In the commercial manufacture of diodes much stronger etches would be used both in the pre- and post-alloy etching stages. A description of one such post-alloy etch which may be used in proper laboratory surroundings to give higher diode yields is given in Sec. A5.5.2.

During the alloying process the Ge wafer surface must remain clean and free from oxide layers. To achieve this, a small amount of solder flux must be applied to the center of the wafer. It is important to limit this quantity severely, and to do this, a fine wire is dipped into the flux and then touched onto the Ge. Either ZnCl† or NH_4Cl in ethylene glycol may be used as a soldering flux. (The NH_4Cl flux does not char so easily, and produces a higher diode yield).

An In sphere 30 mils in diameter‡ now has to be placed on the center of the wafer, in the flux. It is not necessary to clean or etch the sphere, but avoid touching or picking up the sphere with your fingers. A simple way to handle these small spheres is to use a fine wire with a *minute* amount of flux on it. If this wire touches an indium sphere, there is sufficient adhesion to support it, but the sphere can easily be dislodged when it is on the wafer.

The Ge wafer, with the In sphere on it, is now placed on the hot plate surface at a place where the temperature is accurately known to be 200°C (in equilibrium).§ It is extremely important that the actual alloying temperature be known. To accomplish this, the system shown in Fig. A16 may be employed.

It is now best to observe, if possible, precisely when the In melts; the sphere will appear more silvery, or will quiver and sag a little (about 15 to 20 sec after the wafer is placed on the hot plate).

† Commercially this compound is available as Nokorode Soldering Paste, made by M. W. Dunton Co., Providence, R. I.

‡ Available from Indium Corp. of America, 1676 Lincoln Ave., Utica, N. Y. Price is about 1 to 2c per sphere in lots of 500 to 1000.

§ This temperature is much lower than that used commercially (which is near 500°C), and it will yield only a very small amount of Ge-In molten solution at best. The flux used here, the tinning method, and the lack of jig support for the In sphere limit the temperature we can use.

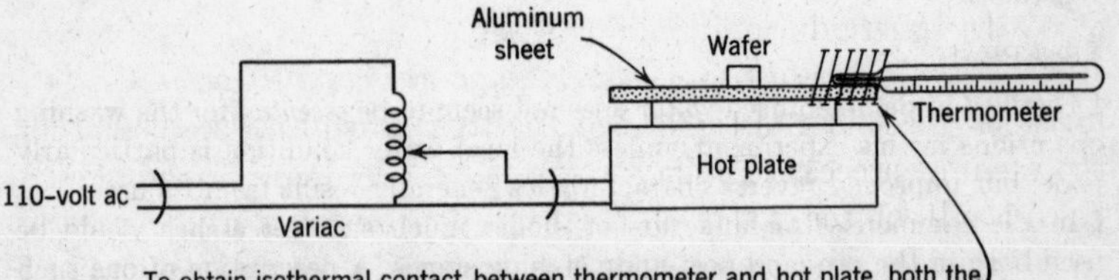

Fig. A16. Determination of alloying temperature.

After these indications, wait an additional 30 sec to allow the temperature to rise above the In melting point, thus producing more of the Ge-In solution, and then remove the wafer and allow it to cool to room temperature. Too much sagging of the In sphere at this point, however, is undesirable, because it may bridge the junction. (Occasionally it is difficult to see when the In has melted. In such cases, 2 min is the maximum time that the wafer should be allowed to remain on the hot plate surface at 200°C. This avoids excessive charring or boiling of the flux.)

The diode is next cleaned by using toluene, acetone, or alcohol, and then rinsed in water.

A5.3 FINAL ETCH AND SURFACE TREATMENT

As indicated in Sec. A5.1.2, the usual functions of a final etch are to:

(1) Remove any alloy dot material which bridges, or even short-circuits, the junction by drooping over it.
(2) Dissolve away enough surrounding material to expose the Ge-In regrowth region at the junction, so it is not close to any of the unalloyed Ge surface.
(3) Remove from the junction surface any residues, tiny fragments, or oxides which result from heating the wafer and the flux.

In the present experiment, however, only the simplest and safest etch is used, namely, immersion of the diode for 2 to 3 min in boiling USP 3% H_2O_2. The diode is then cleaned with alcohol (or toluene or acetone), rinsed with water, and dried. Evidently only function (3) above is performed by such an etch. The other two functions presumably have been performed sufficiently well by the initial surface preparation and care to avoid excessive drooping of the In sphere during the alloy stage.

Next it is necessary to remove the poorly conducting layer that forms on the solder of the tinned (bottom) side of the wafer during the etching process. The layer is removed by lightly sanding the tinned side of the wafer with emery polishing paper. Be careful to insure that the wafer does not become scratched or chipped while being held with the tweezers during sanding. The wafer is now washed again in alcohol and rinsed extremely well in water.

Because any water vapor left on the surface of the wafer after washing decreases the reverse resistance, the diode should be dried thoroughly before testing. Placing the diode (with the In sphere facing down) on a piece of tissue for 15 to 20 minutes will insure adequate drying while the following mount is prepared.

A5.4 DIODE HOLDER

A holder into which the diode may easily be inserted, or from which it may easily be removed, and which also provides electrical contact to the *p* and *n* regions, is illustrated in Fig. A17. Note that the wire used to contact the sphere should be fairly stiff; and before mounting on the clothespin it should be pointed by cutting diagonally, as in Sec. A4.3. The suggested type is a piece of "header" wire, which is a portion of one of the lead wires from a transistor base. On small transistors, these are usually made of Kovar, sometimes gold plated.

A5.5 DIODE CHARACTERISTICS

A5.5.1 *Multimeter Evaluation*

A simple test of the quality of the diode can be made using the resistance range of a multimeter. A forward impedance of a few 10's of ohms should be observed. The reverse impedance depends on

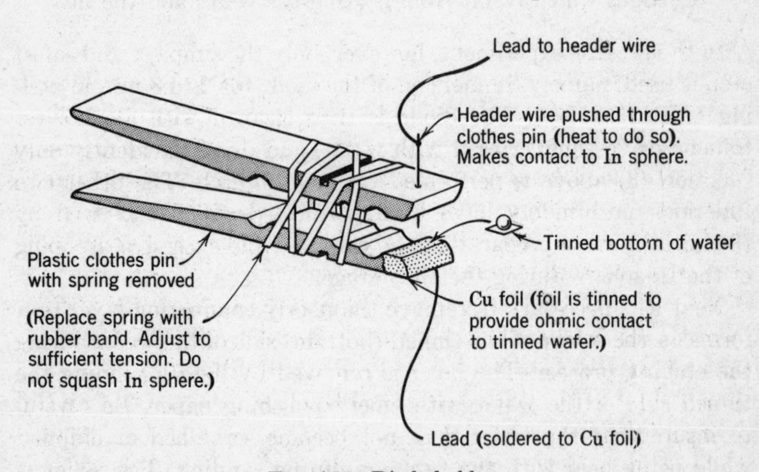

Fig. A17. Diode holder.

the voltage supplied by the meter during the test, but should be of the order of 100 K ohms for a 1.5-volt supply in the meter. (If the multimeter supply voltage exceeds the diode breakdown voltage, no meaningful conclusions can be drawn from this measurement. In this case a transistor curve tracer or ordinary oscilloscope must be used for *rapid* evaluation.

A5.5.2 *Criteria for Success*

The principal features to look for are a low voltage drop at moderate currents in the forward direction, and a fairly clear small reverse "saturation" current prior to reverse "breakdown" (this is where reverse current begins to increase significantly with reverse voltage). A "good" diode made by the present procedure has 50 to 100 ohms "forward resistance," a reverse "saturation current" of a few microamps at about $\frac{1}{2}$ volt reverse bias, and an incremental resistance of around 100 K in that region. A "soft" reverse breakdown usually occurs near 1 volt.

Presented in Table A-1, on page 238, are some indications of faulty diodes and their probable causes.

Only in case (*a*) is a further H_2O_2 etch and subsequent cleaning likely to improve the diode. In all other cases, there are only two choices: start a new diode, or use an etch strong enough actually to remove material from the junction region. These two procedures take about the same amount of time.

A new diode can always be made on the old wafer by simply repolishing the top surface and repeating the procedure from that point on—until the wafer gets too thin!

If a suitable, safe, chemical laboratory environment is available, along with appropriate electrode materials, an electrolytic etch may be employed (see Fig. A18). Whereas this etch may not seem as dangerous as many used commercially, **NaOH AND KOH ARE STRONG BASES AND SHOULD NOT BE USED EXCEPT IN A PROPER SCHOOL OR INDUSTRIAL CHEMICAL LABORATORY, UNDER PROFESSIONAL SUPERVISION, WITH ADEQUATE PROTECTION OF SKIN AND CLOTHING!**

In any case, after each cleanup etch, the tinned bottom of the wafer must be sanded prior to remounting for test. These pro-

Reverse Resistance	Curve Tracer Display (horizontal 0.2 v/cm, vertical 0.1 ma/cm)	What Happened?
(a) 10's of kilohms	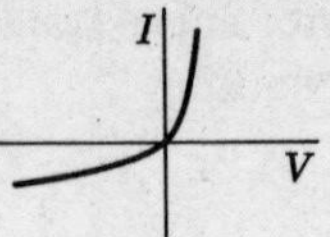 I, V	Ge surface contaminated
(b) 1–10 kilohms	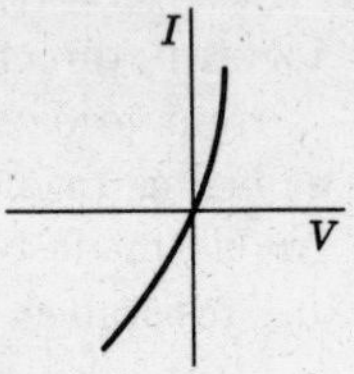 I, V	(1) Hot plate temperature too low → insufficient wetting of In on Ge (2) Hot plate temperature too high → charred flux, leaving Ge surface dirty
(c) 10's–100's of kilohms	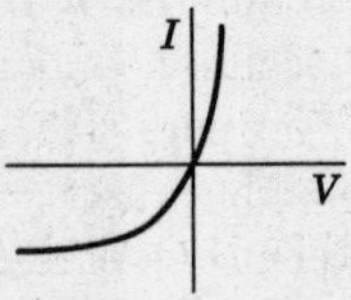 I, V	Ge surface scratched or pitted → high saturation current
(d) 0	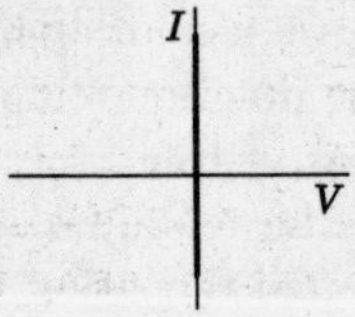 I, V	Hot plate temperature too high (> 215° C), causing excessive droop of indium over junction and resulting short circuit

Table A-1

cedures are repeated until a "good" diode is made. An example appears in Fig. A19. A more careful study of the characteristics is then in order.

A5.5.3 *Diode Volt-Ampere Characteristics*

The circuit of Fig. A20 is suitable for the determination of the diode volt-ampere characteristics at low levels. The voltmeter is placed directly across the diode and its current drain versus voltage reading is first found with the diode absent. The appropriate correction is later made to the current readings when the diode is in the circuit.

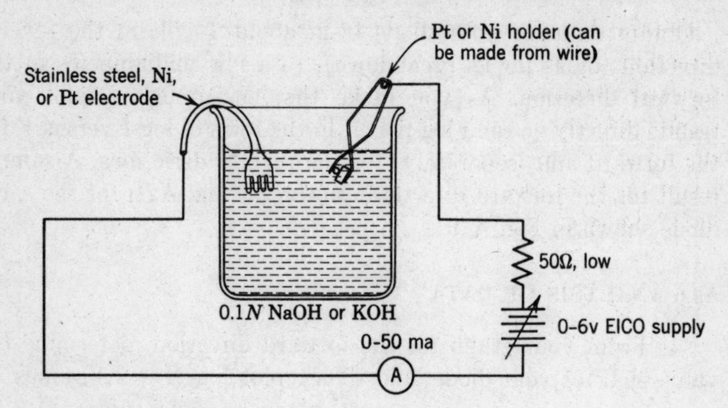

Typical current level for our junction area: 25 ma.
Typical etching time: 1 to 3 min.

Fig. A18. Electrolytic etching arrangement.

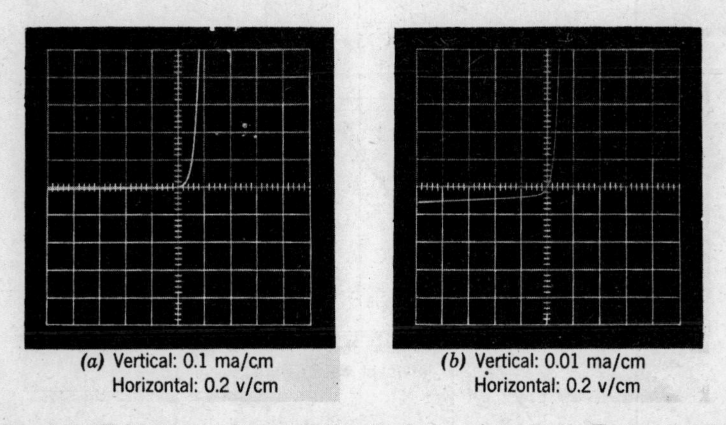

(*a*) Vertical: 0.1 ma/cm
Horizontal: 0.2 v/cm

(*b*) Vertical: 0.01 ma/cm
Horizontal: 0.2 v/cm

Fig. A19. Volt-ampere characteristic of a home-made diode. Forward resistance about 100 ohms; back resistance about 4×10^5 ohms. Diode prepared with H_2O_2 (USP 3%) pre- and post-alloy etch.

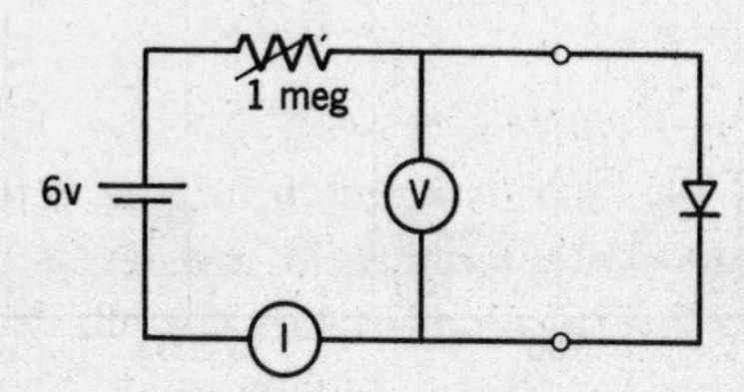

Fig. A20. Circuit for determination of diode volt-ampere characteristic.

Obtain data for your diode from about 1 volt in the reverse direction (or as far as breakdown), to a few milliamperes in the forward direction. As you make the measurements, plot your results directly on semi-log paper, in the form of log I versus V for the forward and (separately) for the reverse directions. A sample result for the forward direction appears in Fig. A21, for the same diode shown in Fig. A19.

A5.6 ANALYSIS OF DATA

1. From your graph for the forward direction, determine the value of I_s for your diode, and then replot log $(I + I_s)$ versus V.

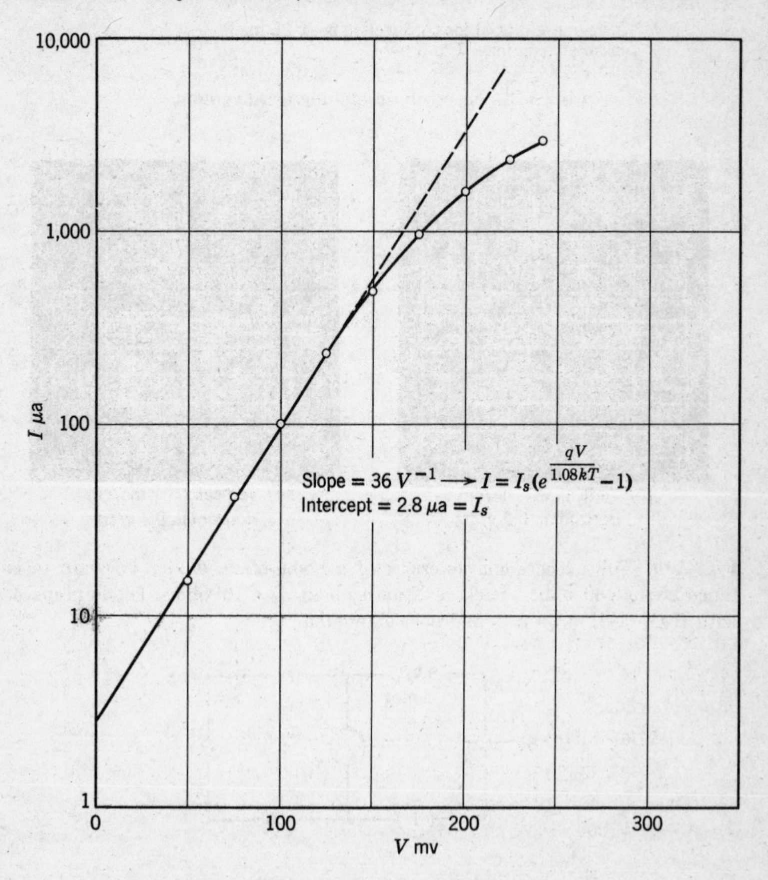

Fig. A21. Characteristic of home-made alloy diode.

Comment on the fit of experimental points at very low currents on the original curve and on the new one. Note particularly the strong effect of stated ammeter and voltmeter errors at readings which are made at small fractions of full scale.

2. Estimate I_s from the reverse part of your volt-ampere characteristic, and compare with the previous value. Comment on possible reasons for difference. Consider using more accurate meters, if available.

3. Determine from the forward characteristic whether the exponential voltage dependence agrees quantitatively with the simple theory of Eq. A1. Again, consider the effect of meter errors.

4. How do you account for the behavior at large forward currents? Make a quantitative estimate of this effect, and compare with what you predict from the diode geometry and the resistivity of the wafer.

5. Make at least one more "good" diode, and note carefully how the results differ from the first one.

6. If the diode is placed in its holder with the header wire contact resting directly on the Ge surface rather than on the In dot, we are able to observe a diode characteristic (do it). Vary the point pressure and note the effect. Study one case by the same methods used for the previous diode. Describe the main differences between the two characteristics. Move the point to another spot and repeat. Observe carefully any differences in results.

7. Try some different kinds of wire in this experiment and *also* as contacts for the In dot on your alloy diode. Try *carefully* to change contact pressure on the In dot, without dislodging it! What do your results suggest?

8. Try "forming" your point-contact diodes in the manner of Sec. A4.3, and examine the effect on the volt-ampere characteristics. You *may* need to try a range of capacitor and charging-voltage values to find an effective combination for your wafer and type of wire. Summarize the effect of "forming" on the *V-I* characteristic.

9. What happens if you "form" your alloy diode?

10. What is suggested about the difference in nature between the two types of junctions, by your observations of the "forming" action on both types? Is this consistent with your observations of the effects of changing pressure and type of wire? How about the "repeatability" of both types of junction?

Index